Changing the Faces of Mathematics

Perspectives on Asian Americans and Pacific Islanders

Series Editor

Walter G. Secada
University of Wisconsin—Madison
Madison, Wisconsin

Editor

Carol A. Edwards
Saint Louis Community College
at Florissant Valley, Missouri
(retired)

National Council of Teachers of Mathematics
Reston, Virginia

Library of Congress Cataloging-in-Publication Data

Perspectives on Asian Americans and Pacific Islanders / editor, Carol A. Edwards.
    p. cm. — (Changing the faces of mathematics)
  Includes bibliographical references.
  ISBN 0-87353-475-1
    1. Mathematics—Study and teaching—United States. 2. Asian American
students—Education. 3. Pacific Islander American students—Education. I. Title: Asian
Americans and Pacific Islanders. II. Series. III. Edwards, Carol A.

QA13 .P48 1999
510′.71′073 21—dc21

                                    99-045435

Printed in the United States of America

Contents

1. Explaining Mathematics through Asian Folktales . . . 1

Denisse R. Thompson
University of South Florida, Tampa, Florida

Michaele F. Chappell
University of South Florida, Tampa, Florida

Richard A. Austin
University of South Florida, Tampa, Florida

2. Hmong Needlework and Mathematics 13

Joan Cohen Jones
Eastern Michigan University, Ypsilanti, Michigan

3. Using Asian-Pacific Literature to Enhance
Problem Solving in the Elementary School
Mathematics Classroom 21

Nancy C. Whitman
University of Hawaii, Honolulu, Hawaii

4. Differences in Arithmetic Nomenclature 29

Jean Bee Chan
Sonoma State University, Rohnert Park, California

5. Counting in Chinese, Japanese, and Korean:
Support for Number Understanding 31

Irene T. Miura
San Jose State University, San Jose, California

Yukari Okamoto
University of California, Santa Barbara, California

6. Connecting Mathematics, Language, and
Culture for ESL Students: A Guam Classroom 39

Jacquelyn Milman
University of Guam, Mangilao, Guam

Ione Wolf
University of Guam, Mangilao, Guam

Wilson Tam
Guam Department of Education, Agana, Guam

Preface

In 1993, the National Council of Teachers of Mathematics created a task force on multiculturalism and gender in mathematics. That task force recommended the publication of a six-volume series that would help make the slogans of "mathematics for all" and "everybody counts" realities. The Council approved that recommendation.

In 1994, an editorial panel for Changing the Faces of Mathematics, as this series is called, was constituted to oversee the preparation of those volumes, each with a different focus but also with important commonalities. The Editorial Panel looked for manuscripts that were at the intersection of research and practice, that is, manuscripts that said that something works and that also provided reasoned explanation about the principles that underlie why that "something" works. The panel also looked for manuscripts that spoke clearly to many audiences—teachers, curriculum supervisors, department chairs, principals, program developers, and administrators. Finally, the panel was interested in manuscripts that moved beyond single-dimensional analyses of equity issues.

This book contains eight manuscripts that address the teaching of mathematics to Asian and Pacific Island children and the roles of language and culture in the classroom. Contributors are from Florida to Guam, bringing multiple perspectives to readers and reflecting the Editorial Panel's desire that classroom teachers contribute to this series.

Manuscripts by Thompson, Chappell, and Austin; by Jones; and by Whitman are examples of the rich traditions and contributions of Asians and Pacific Islanders to mathematics. They provide opportunities for discussions and explorations of the mathematics and background of Asian and Pacific Island culture using cultural artifacts and literature as a springboard and basis for introducing mathematical ideas in the classroom.

Manuscripts by Chan and by Miura and Okamoto focus on language. Understanding that some Asian children may have been taught to read some mathematical notations differently from most children in the United States will help make all teachers and students sensitive to such differences.

The article by Milman, Wolf, and Tam segues from a focus on language to a focus on culture since the article looks at both these elements. The authors discuss cultural influences on the classroom structure as well as the usage of language.

The final group of manuscripts by Ahuja, Matsushita, and Randolph and by Dougherty, Slovin, and Matsumoto relate to cultural differences that influence classroom dynamics, classroom behavior, and environment. For example, some Asian children may have learned in their cultures to place different levels of significance on exactness in measurement. Expressions and mannerisms in the relationship of children to adults may also be different among Asian children.

Although some of these chapters contain examples that are grade-level specific, they all contain material that will be useful at all levels. The articles should help all readers develop a deeper understanding of, become more sensitive to, and stimulate a desire to learn more about Asian and Pacific Island students and their unique characteristics.

Carol A. Edwards
Editor

Acknowledgments

I would like to acknowledge the help of many people in bringing this volume to fruition. Their help was invaluable. My sincere thanks go to the following people and groups:

- The Editorial Panel of the series, Changing the Faces of Mathematics, for developing the guidelines and for its encouragement and assistance at various stages. Special thanks go to the series editor, Walter Secada, University of Wisconsin—Madison, who reviewed all the manuscripts for this volume.

- The Editorial Panel for this volume, who contacted potential authors, attended meetings, reviewed the submitted manuscripts, and made thoughtful suggestions for improvements:

  Kil S. Lee
  Minnesota State University—Mankato
  Mankato, Minnesota

  K. C. Ng
  California State University
  Sacramento, California

- Manuscript reviewers who offered many helpful suggestions:

  Don Balka
  Saint Mary's College
  Notre Dame, Indiana

  Alan Chan
  Woburn Collegiate Institute
  Scarborough, Ontario

  Roy Winstead
  Brigham Young University—Hawaii
  Laie, Hawaii

- Authors of manuscripts for their interest in this project and patience during the long review and publication process

- NCTM Headquarters Publications Department editorial and production staffs, especially Harry Tunis, who patiently answered many questions.

- My husband, Roger Edwards, who read some of the manuscripts and provided encouragement.

Carol A. Edwards
Saint Louis Community College
at Florissant Valley, Missouri
(retired)

Exploring Mathematics through Asian Folktales

Denisse R. Thompson

Michaele F. Chappell

Richard A. Austin

As teachers strive to make mathematics relevant to the lives and cultures of students from diverse backgrounds, it is essential that they become aware of a variety of resources that facilitate the inclusion of multicultural content in the mathematics classroom. These resources must not only provide exposure to the contributions and cultures of other groups but must also provide opportunities for rich discussions and explorations of mathematics. One such resource is that of Asian folktales.

A number of sources are available that describe Asian folktales in a manner designed to motivate students (Suyeoka, Goodman, and Spicer 1974; Pittman 1986; Birch 1988; Snyder 1988; Howell 1990; Hong 1993; Barry 1994). Athough many examples of connections between literature and mathematics exist at the elementary school level (Burns 1992; Whitin and Wilde 1992, 1995; Zaslavsky 1996), not as many examples exist that illustrate the use of literature at the middle school level. Yet, Sullivan and Nielsen (1996) and Austin and Thompson (1997) show how the use of literature in middle school can enhance students' attitudes toward mathematics by giving students the opportunity to explore mathematical concepts within an encouraging and nonthreatening environment.

To illustrate the possibilities for mathematical study, we have taken one specific folktale, *A Grain of Rice* (Pittman 1986), and developed three lessons addressing different mathematical concepts. After discussing the mathematical possibilities with this folktale, we describe several other folktales with the same basic story that could be used instead. Finally, we list several different folktales and provide a brief sketch of possible mathematical activities that could be used with each story. Although the mathematical concepts explored through any of these folktales do not necessarily depend on Asian culture, the folktales themselves have a strong cultural reference. As with stories from many cultures, the folktales often have a moral in addition to the mathematics. In particular, perseverance, honor, cleverness, and knowledge of mathematics enable the hero of the folktale to achieve the goal; these are useful values for middle school students to consider as they explore mathematics.

SUMMARY OF THE FOLKTALE *A GRAIN OF RICE*

Pong Lo, a peasant farmer, arrives at the emperor's court seeking the hand of the emperor's daughter in marriage. Pong Lo's request makes the emperor very angry. After his rage settles, the emperor decides to give Pong Lo a job in the palace rather than chop off his head. Through several different jobs, Pong Lo demonstrates his cleverness in a variety of ways. When the princess becomes ill and close to death, Pong Lo prepares a potion that provides a cure. Asked by the emperor to name a reward, Pong Lo again requests the hand of the princess in marriage. The emperor refuses and demands that Pong Lo choose another reward. After some thought, Pong Lo requests a single grain of rice, with the amount to be doubled each day for one hundred days. Although the emperor laughs at the request, he agrees to the strange reward. Each day the grains of rice are delivered using an elaborate gift container. After 40 days the emperor realizes that the country does not have enough rice to satisfy the request, at which time Pong Lo has become a rich man. At this point, the emperor permits Pong Lo to marry his daughter, and the request for grains of rice is suspended.

ACTIVITY SHEET 1

The Peasant Problem Solver (Algebraic Thinking)

Activity Sheet 1 on pages 7–9 gives students an opportunity to explore the mathematical patterns described in the story. Although the folktale only focuses on the number of grains rewarded on a given day, it is instructive to have students consider the total number of grains of rice rewarded since the first day. Completion of this table provides an excellent opportunity for students to use appropriate technology, specifically calculator technology, to explore mathematics.

Several opportunities for mathematical discussion occur while completing this table. First, students can discuss different methods of obtaining the entries in the table. Some students may multiply the number of grains rewarded on day n by 2 to find the number of grains rewarded on day $n + 1$. Others may choose to add the number of grains rewarded on day n to itself. Hence, students have the opportunity to observe that two different strategies lead to the same solution.

Second, scientific notation becomes a natural topic of discussion as students complete the values in the table. Regardless of the type of calculator used, at some point the number of grains of rice is displayed using scientific notation. (On two different calculators we used, one scientific and one graphing, the number of grains of rice is recorded using scientific notation on day 35.) Hence, through use of this folktale, students are exposed to scientific notation in a meaningful situation.

Third, the opportunity to graph the results from the table offers the students experience with a nonlinear graph. Although the activity sheet only asks for a graph of the number of grains per day, students can also graph the total number of grains and compare the results. Both graphs are exponential and the graphs provide a visual image of the power of doubling. For students who have studied exponents, this is an appropriate time to consider rules for the number of grains of rice rewarded on day n, namely 2^{n-1}, and the total number of grains of rice $2^n - 1$. Observe that to determine these rules, students need to rewrite the number of grains as a power of 2 in order to relate the exponent to the number of the day.

Depending on the class, teachers can provide an additional extension and contrast by having students graph the number of grains on each day and the total number of grains if Pong Lo received just one grain of rice per day. In this case, the number of grains rewarded on each day will be 1, and students can see a real example of a constant function. For day n, the total number of grains rewarded since day 1 is now just n, and students have a real example of a linear function with slope 1. This extension helps students distinguish between a constant increase (resulting in a linear function) and a constant *rate* of increase (resulting in an exponential function). By having students explore these ideas informally through the folktale, they are in a better position to study these topics abstractly in a formal course in algebra.

As indicated on the activity sheet, students can also compare the number of grains rewarded under a doubling rule with the number rewarded if the emperor had tripled the results each day. Graphing the results from a tripling rule provides another means of illustrating just how much faster the number of grains of rice grows when tripled as compared to doubled.

By comparing the number of grains of rice in a typical bag of rice and the weight of the bag, students can obtain an estimate of the weight of rice on each of the days in the table. Students should keep in mind, however, that the table only contains entries for the first forty days—the number of days until the emperor changed his mind about the marriage. Originally, Pong Lo was to receive rice for one hundred days.

Storing a Fortune (Volume Measurements)

The second activity sheet on page 10 offers students an opportunity to explore volume while considering gifts with cultural significance. Each of the gifts used by the emperor to deliver the grains of rice was an item that was valued within the Chinese culture. Some of the items may be quite unfamiliar to students; in such cases, pictures in a book or from catalogs can be used to help students visualize the gifts.

To estimate the dimensions of a box capable of holding the gift container, students must consider the size needed to hold the grains of rice on a given day. Hence, this activity forces students to interpret the information from Activity Sheet 1 in a new manner. Students must find an estimate for the volume of the rice on a given day. Many techniques are possible for estimating this volume (Welchman-Tischler 1992). For instance, students may determine the volume for a given number of grains and then use this result to estimate the volume of the rice for a specific day. Then they can connect the volume of the rice to the volume of a box containing the emperor's gift. Because many different box dimensions yield the same volume, and many different sets of dimensions are valid, students should justify a set of dimensions that are reasonable.

In determining whether or not all the gift containers would fit in the classroom, students must not only consider the dimensions of the classroom but must also realize that many days are not represented in the table. Students can determine their own choice of a gift container for the missing days and give some justification for their choice.

If resources such as catalogs are available, or if import-export stores exist in the area, a natural extension is to consider the monetary value of each container. After all, the accumulation of these gifts led to Pong Lo's rise from poverty to riches. On several days, duplicates of a particular gift were needed to deliver all the rice. Hence, students can consider a unit price for the gift and then determine the value of all the duplicates of that gift needed on that day.

ACTIVITY SHEET 3

Preparing a Feast (Problem Solving, Numerical Operations, and Statistics)

This final activity sheet (page 11) gives students an opportunity to plan a banquet, comparing and contrasting costs from more than one restaurant. If students live in a location with an Asian grocery store, they can compare the costs of a restaurant banquet with the costs of preparing the entire meal from scratch.

An additional means to enhance the problem-solving aspect of this task is to give students a fixed amount of money for the banquet. Students are expected to come as close as possible to the given amount without overspending. This extension leads students to consider the cost of each particular item on the overall expense as they attempt to use as much of the money as possible.

All of the activities on this sheet are good activities for students to use in a cooperative problem-solving environment. Different students within each group could investigate costs and uses of different varieties of rice, with all students pooling their investigations to write a final report. Further, question three can be extended by having different groups of students investigate the production of rice in different countries. The findings from the various groups may result in an overall class display that contains information about all the world's rice-producing countries.

OTHER VERSIONS OF THIS FOLKTALE

Other variations of this folktale can also be used to explore mathematics. *The King's Chessboard* (Birch 1988) and *The Rajah's Rice* (Barry 1994) are Indian versions of this folktale in which a servant performs a service for the ruler and is rewarded with grains of rice on a chessboard. In *The King's Chessboard*, one grain of rice is placed on the first square of a chessboard on the first day; two grains are placed on the second square on the second day; four grains are placed on the third square on the third day; and so on, with the number of grains doubling each day. In *The Rajah's Rice*, two grains of rice are placed on the first square, with the number of grains doubling for each subsequent square. In both tales, students have the opportunity to explore the same mathematical patterns as those in *A Grain of Rice*. The slight variation in these tales provides a nice contrast that shows how a minor change in the beginning of the problem has major implications at the end, as the number of grains on the last square of the chessboard in the first tale is 9.22×10^{18}, whereas it is 1.84×10^{19} for the second tale.

OTHER ASIAN FOLKTALES

Several other Asian folktales exist that are appropriate for investigating middle school mathematics. *Two of Everything* (Hong 1993) is a Chinese folktale involving the same doubling pattern as the previous examples, but it is set in a different context. In this tale, an elderly gentleman finds a brass pot that doubles everything put inside, including people. After the gentleman and his wife accidentally produce copies of themselves, they use the brass pot to create two of everything they have so that both couples can live happily ever after. Students could investigate how long it would take a given amount of money to reach a specified level if the pot doubles, triples, or takes half of the amount placed inside.

Issunboshi (Suyeoka, Goodman, and Spicer 1974) describes the adventures of a Japanese boy just one-inch tall. After reaching the age of manhood, he leaves his parents' house to enter the service of the emperor and becomes a protector to the princess. When he saves the princess from attack, he is rewarded with a wish, to become a man-sized warrior. Numerous opportunities exist for exploring proportions and concepts of relative size.

The Story of Chinaman's Hat (Howell 1990) is a folktale with a Chinese-Hawaiian connection that describes the origin of a small island off Oahu known as Chinaman's Hat. After drinking too much of a magic potion, a boy grows enormously tall and is able to float from China to Hawaii. The concepts of proportion and relative size are among the many mathematical concepts that can be studied through this story. Further, the distance from China to Hawaii can be used to consider travel times—by boat, by plane, and by swimming. As part of this investigation, students can consider the difficulties many Pacific Islanders experienced as they traveled from island to island throughout the Pacific.

Finally, *The Boy of the Three-Year Nap* (Snyder 1988) offers a Japanese tale in which a lazy boy uses problem solving to win the hand of a wife. His mother also uses clever problem solving to improve the condition of her home and change her son into an industrious individual. Here, students can investigate different home construction techniques in different parts of the world, as well as construction costs. As part of this exploration, students can consider some of the ancient buildings and structures that exist in many Asian cultures and how the designs have withstood natural disasters over centuries.

CONCLUSION

Because students of all ages enjoy stories, literature provides a motivating vehicle for exploring mathematics. Folktales, in particular, generally portray positive aspects of a society and provide a cultural context in which to deliver a strong moral message. Within that cultural context, the folktale is a means of exploring mathematics in an environment that captures students' attention. The mathematics that is studied through a folktale depends solely on the needs of the teachers and the students. We encourage teachers to consider using such folktales with their students throughout the school year as they study mathematics. It is a natural way to infuse multicultural connections into the classroom.

REFERENCES

Austin, Richard A., and Denisse R. Thompson. "Exploring Algebraic Patterns through Literature." *Mathematics Teaching in the Middle School* 2 (February 1997): 274–81.

Barry, David. *The Rajah's Rice: A Mathematical Folktale from India.* New York: W. H. Freeman and Company, 1994.

Birch, David. *The King's Chessboard.* New York: Puffin Pied Piper, 1988.

Burns, Marilyn. *Math and Literature (K–3): Book One.* Sausalito, Calif.: Math Solutions Publications, 1992.

Encyclopaedia Brittanica. "Forbidden City." *The New Encyclopaedia Britannica (15th Edition).* Chicago: Encyclopaedia Britannica, 1995.

Hong, Lily Toy. *Two of Everything.* Morton Grove, Ill.: Albert Whitman and Company, 1993.

Howell, Dean. *The Story of Chinaman's Hat.* Honolulu, Hawaii: Island Heritage Publishing, 1990.

Pittman, Helena Clare. *A Grain of Rice.* New York: Bantam Skylark, 1986.

Snyder, Dianne. *The Boy of the Three-Year Nap*. Boston: Houghton Mifflin Company, 1988.

Sullivan, Emilie P., and William Nielsen. "Fictional Literature in Mathematics." *Mathematics Teaching in the Middle School* 1 (January-February 1996): 646–47.

Suyeoka, George, Robert B. Goodman, and Robert A. Spicer. *Issunboshi*. Aiea, Hawaii: Island Heritage Publishing, 1974.

Welchman-Tischler, Rosamond. *How to Use Children's Literature to Teach Mathematics*. Reston, Va.: National Council of Teachers of Mathematics, 1992.

Whitin, David J., and Sandra Wilde. *It's the Story That Counts: More Children's Books for Mathematical Learning, K–6*. Portsmouth, N.H.: Heinemann, 1995.

———. *Read Any Good Math Lately? Children's Books for Mathematical Learning, K–6*. Portsmouth, N.H.: Heinemann, 1992.

Zaslavsky, Claudia. *The Multicultural Math Classroom: Bringing in the World*. Portsmouth, N.H.: Heinemann, 1996.

ACTIVITY SHEET 1: THE PEASANT PROBLEM SOLVER

Read the story *A Grain of Rice* by Helena Clare Pittman. In the story, Pong Lo is supposed to receive a reward of rice from the emperor for 100 days. The table below has places to show the number of grains of rice to be rewarded on each of the first forty days as well as places to record the total number of grains of rice rewarded. Fill in the columns of the table. Some entries from the story are given; use these to check your work as you complete the table. Then use the table to answer the questions.

Day	Number of Grains Rewarded on This Day	Total Number of Grains Rewarded Since Day 1
1	1	1
2	2	3
3	4	7
4	8	
5	16	
6	32	
7	64	
8	256	
9	512	
10		
11		
12	2048	
13		
14		
15		
16		
17		
18	131 072	
19		
20	524 288	
21		
22		
23		
24		
25	16 777 216	
26		
27		
28		
29		
30	536 870 912	
31		
32		
33		
34		
35		
36		
37		
38		
39		
40	549 755 813 888	

1. a. What is the first day for which Pong Lo's reward is more than a thousand grains of rice?

    b. What is the first day for which Pong Lo's reward is more than a million grains of rice?

    c. What is the first day for which the TOTAL number of grains of rice is more than a thousand grains? More than a million grains?

2. Use the coordinate graph below to graph the number of grains of rice rewarded on each of the first ten days.

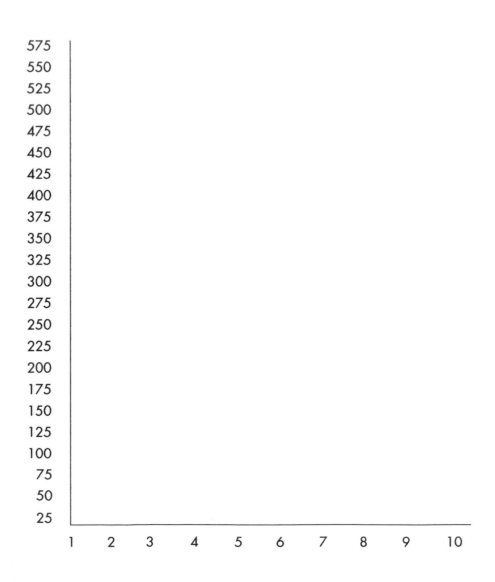

3. One way to find the number of grains rewarded on the hundredth day is to continue the table. Another way is to find a rule that gives the number of grains.

   a. Use the information in the table to find a rule for the number of grains of rice on any day n.

   b. Use your rule to find the number of grains rewarded on the 100th day.

4. Suppose the emperor tripled the number of grains of rice each day, instead of doubling.

   a. What would be the first day for which Pong Lo would receive more than a thousand grains of rice?

   b. What would be the first day for which Pong Lo would receive more than a million grains of rice?

5. a. Find a bag of rice. Record its weight and then count the number of grains of rice in the bag.

   b. Use the results from 5 (a) and the table to estimate how much the rice would weigh on day 10, day 20, and day 30.

ACTIVITY SHEET 2: STORING A FORTUNE

The emperor delivered the rice to Pong Lo using elaborate containers worth a great deal of money. The table below describes some of the containers the emperor used to deliver the rice. Suppose each container is put in a box just large enough to hold it. Estimate the height, length, and width of the box; keep in mind that the container and the box must be large enough to hold the grains of rice rewarded on that day. Determine the volume of the box holding each container.

Day	Gift	Length of Box	Width of Box	Height of Box	Volume of Box
1	silver bird nest				
2	china cup				
3	alabaster swan				
4	enameled bowl				
5	golden plate				
6	dragonfish				
7	boat of precious stones				
8	jade box				
9	ivory tray				
12	embroidered silk box				
18	four ebony chests				
25	25 brocaded sacks				
30	40 brass urns				
40	100 rosewood and mother-of-pearl trunks				

1. Consider the measurements of the boxes in the table.
   a. How much room would Pong Lo need to store all the gifts if they were placed in the appropriate boxes?

   b. Find the measurements of your classroom. How many classrooms would be needed to hold all of Pong Lo's gifts? Justify your answer.

   c. Which of the gifts in the table could fit in your classroom? Justify your answer.

2. The Forbidden City, a complex of the former imperial palaces, has dimensions 2.5 miles by 2.5 miles by 35 feet high. Is the Forbidden City large enough to hold all of the gifts in the table? Justify your answer.

3. Suppose you were responsible for storing all of Pong Lo's gifts. Design a palace for holding the gifts and a plan for arranging them within the palace. Make the palace large enough to live in as well as to store the gifts. Describe your plan with appropriate diagrams.

ACTIVITY SHEET 3: PREPARING A FEAST

1. One of Pong Lo's many jobs involved helping prepare the royal meals. Obtain menus from one or more Asian restaurants in your vicinity.

   a. Suppose your class decides to hold an end-of-the-year banquet featuring Asian cuisine. Plan an appropriate menu for your class and determine the amount of food needed.

   b. Use your menus to compare the cost of the banquet from several different restaurants. Which one gives the better value? Justify your answer.

2. There are many different varieties of rice. Find out about some of these varieties. In particular, find some of the different uses and differences in costs among the varieties. Write a short report describing your findings. Use graphs when appropriate to display your results.

3. Use an almanac or an encyclopedia to find the top five rice-producing countries in the world.

   a. Based on the amount of rice they produce each year, estimate how many grains of rice they produce.

   b. Consider the population of each of the countries. How much rice do they produce per person?

   c. Would any of the countries produce enough rice to reward Pong Lo? Why or why not?

   d. Try to find the total rice production in the world. Is it enough to reward Pong Lo for all one hundred days? Justify your answer.

Hmong Needlework and Mathematics

2

Joan Cohen Jones

The Hmong came to Wisconsin, Minnesota, California, Georgia, and other states as refugees from Southeast Asia. In an effort to understand and respect the culture of Hmong children and their families and to make learning relevant to what Hmong children already know, elementary teachers in a small midwestern city have developed interdisciplinary lessons motivated by the exquisitely detailed Hmong needlework. These lessons contain significant mathematical content and are aligned with the NCTM *Professional Standards for Teaching Mathematics*, which call for teachers to understand and use in their teaching "the role of mathematics in culture and society" (National Council of Teachers of Mathematics 1991, p. 132).

A majority of the Hmong children that we encounter in the elementary schools are first-generation immigrants. In most instances, their parents speak little or no English. Their home life, values, and culture are disconnected from what they learn in the classroom. These lessons were developed to help Hmong students perform well in mathematics and other disciplines "by using their cultural backgrounds as a pedagogical resource" (Sleeter 1997, p. 682). For Hmong students who have little knowledge of their own cultural heritage, the lessons help them discover their own history and culture. The teachers who use these lessons report positive results for all their students. Multicultural curricula help build understanding and acceptance of different cultures. In particular, multicultural mathematics makes mathematics more interesting and helps children understand how mathematics was created and is used by people all over the world.

These lessons were designed to encourage discussion, conjecture, and active learning. They are best implemented with students working cooperatively in small groups. As Sleeter (1997) reports, cooperative group learning in mathematics benefits culturally and ethnically diverse students, especially students for whom English is a second language.

Included here is information about the history of the Hmong, the cultural significance of their needlework, the mathematics of their needlework, ideas for interdisciplinary lessons, and connections to other cultures.

A PEOPLE IN EXILE

During the Vietnam War, the Hmong helped the American side by rescuing American pilots, attacking enemy supplies, and gathering valuable information. Many Hmong were killed. After the war, about 300 000 Hmong sought refuge. Many fled to Thailand, where they lived and still live in refugee camps. By 1990, approximately 90 000 Hmong had emigrated to the United States.

HMONG NEEDLEWORK

Paj ntaub (translated "flower cloth"), pronounced "pon dow," is the traditional needlework of the Hmong. Designs are created on cloth in a variety of ways, including batik, reverse applique, and embroidery. Reverse applique uses two layers of cloth with designs cut in the top layer so that the bottom layer shows through. The two layers are then joined with tiny stitches and embroidery. Historically, Hmong girls learned to sew at an early age, usually seven or eight. Their mothers and grandmothers taught them the necessary skills. Patterns were never written down but were learned from watching others. And measurements were never taken. Girls were instructed to make stitches as long as a grain of rice (Shea 1995). Skillful needlework was admired by all and helped raise the prospects of future brides. Today, not all Hmong girls are taught to embroider the paj ntaub and, because of economic necessity, some men have begun to embroider.

The preparation for a paj ntaub is quite mathematical. The sewer divides the cloth into four equal quadrants by making lines in the cloth with a needle. Next, a grid of equally spaced vertical and horizontal lines is made by the same method. Then the cloth is ready for sewing. Traditional designs for the paj ntaub are drawn from nature and include "elephant foot," "snail," and "dragon tail."

After the Hmong arrived in the refugee camps of Thailand, the women created a new kind of embroidery. Called *story cloth*, each consists of a large piece of cloth with embroidered people, animals, trees, and villages. Each story cloth provides a visual record, depicting traditional village life, Hmong folktales, or escape from Laos. Each cloth has a border of triangles, which some say represent the mountains of Laos. Most story cloths are made of blue fabric with gray triangles for the border. Story cloths were created by Hmong women to provide a historical record for their children. The Hmong had no written language before coming to the United States and chose embroidery as a means of recording and communicating their history. Some say that story cloths at first contained hidden messages that were used for espionage. No verfication of this has been established.

While the Hmong lived in refugee camps in Thailand, their embroidery caught the attention of traders, who suggested that Hmong women embroider larger patterns. The Hmong like very bright colors, but traders suggested that paj ntaub be created in softer hues, more to the liking of western customers. Today, Hmong embroidery has become quite valuable. Hmong families sell their embroidery at craft fairs and in small shops. An excellent resource for Hmong designs can be found in *Hmong Textile Designs* by Anthony Chan (1990).

IDEAS FOR MATHEMATICS LESSONS

Most Hmong designs contain horizontal and vertical line symmetry, and some contain rotational symmetry. Students can investigate mathematics within the context of creating a paj ntaub. *The Hmong in America* by Peter and Connie Roop (1990) provides instructions for making a paper paj ntaub, which we have modified to facilitate the mathematics.

Objectives

1. Students design a paj ntaub with line or rotational symmetry or both.
2. Students determine their own definitions of rotational and line symmetry.
3. Extensions: Students find the perimeter and area of squares and examine the similarity properties of polygons.

Materials

Each student needs two squares of different-colored construction paper (sized 8″ × 8″ and 10″ × 10″) and one piece of 6″ × 6″ tracing paper. A piece of cardboard and some extra construction paper are helpful.

Making the Paj Ntaub

1. Fold the 6″ × 6″ square vertically and horizontally, so that it has four equal-sized quadrants (see fig. 2.1a).
2. Starting in the upper left quadrant and continuing clockwise around the figure, label each quadrant with the numbers 1, 2, 3, and 4 in pencil (see fig. 2.1b). The quadrant with number 1 in the upper left-hand corner will be referred to as *quadrant 1*.
3. Select a design from figure 2.2 on page 16 and trace this design once in each quadrant of the 6″ × 6″ square. There are two methods of tracing that will be described below.

The trace-and-turn method

1. Trace the selected design in quadrant 1 (see fig. 2.1c).
2. Turn paper counterclockwise so that quadrant 2 is in the upper left position. Trace the design again (see fig. 2.1d).
3. Continue in this way until you have traced the design in quadrant 3 (see figure 2.1e) and quadrant 4 (see fig. 2.1f). Does your design, as a whole, look symmetric? Will this method work for all designs?

The fold-and-trace method

1. Trace the selected design in quadrant 1 (see fig. 2.3b on page 17).
2. Fold square along vertical axis, so that quadrant 1 is on top of quadrant 2 (see fig. 2.3c).
3. Flip paper so that quadrant 2 is on top of quadrant 1 with your drawing facing down (see fig. 2.3d). The number 2 should be visible (make sure the front side of quadrant 2 is facing up).
4. You should be able to see the design from quadrant 1 beneath quadrant 2. Trace the design.
5. Open square so that all four quadrants are visible (see fig. 2.3e).
6. Fold along horizontal axis so that quadrant 1 is on top of quadrant 4 (see fig. 2.3f).
7. Flip, so that quadrant 1 faces down (see fig. 2.3g). Trace again.
8. While the paper is folded horizontally, trace the design in quadrant 3 from the design beneath it in quadrant 2 (see fig. 2.3h).
9. Open the square and check the resulting design for symmetry (see fig.2.3i).

Suggested Small Group Activities

1. Demonstrate both methods with individual designs. As an example, the "worm track" can be reproduced using the fold-and-trace method, but not the trace-and-turn method, because the design itself does not have rotational symmetry. The "tick," however, can be reproduced by either method because the tick does have rotational symmetry.
2. Give each group two or three Hmong designs from figure 2.2. Ask each group of students to trace their designs by using either of the tracing methods described above, one at a time, in each of the four quadrants of the 6″ × 6″ square of tracing paper so that the completed paj ntaub (see fig. 2.4)

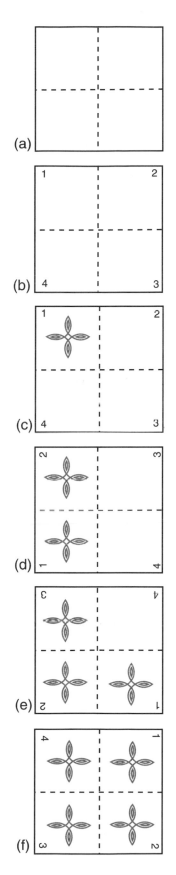

Fig. 2.1. Trace-and-turn method

Worm Track

Tiger Eyebrow

Fish Scales

Spider Web

Elephant's Foot

Horns

Cucumber Seed

Tick

Rat Eye

Snails

Ram's Head

Heart

Fig. 2.2. Hmong designs

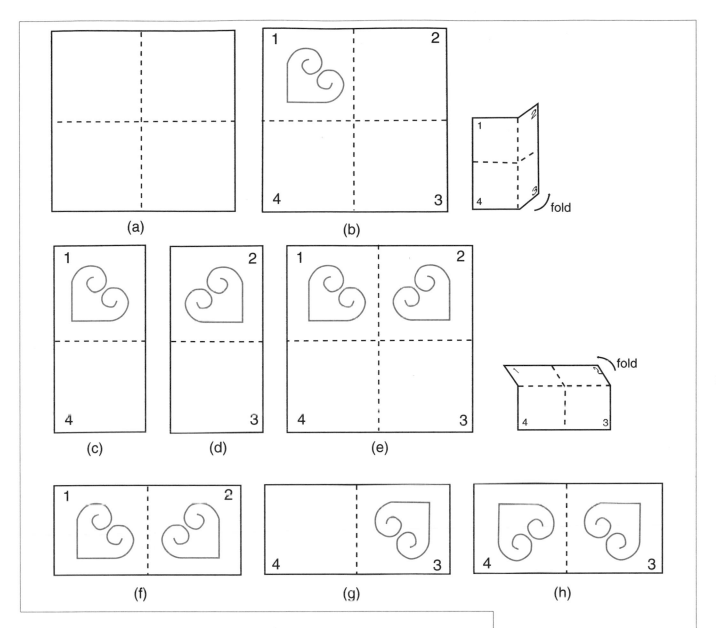

(a)

(b)

(c)

(d)

(e)

(f)

(g)

(h)

has horizontal and vertical symmetry. Ask students to keep track of which tracing methods they used for each design

3 After students have completed tracing their designs, ask different groups to share their results. Ask: How do we know whether the paj ntaub has line symmetry? How can we check? Do any of the paj ntaub look the same after they are turned? What type of symmetry is this?

4. Classify individual designs like "heart," "rat eye," and "cucumber seed" by asking which tracing method worked for each one. Sort the designs into two groups, those that used the trace-and-turn method and those that used the fold-and-trace method. Are individual designs in both groups? What do you know about a design if the trace-and-turn method worked?

5. We used the fold-and-trace method to create a design that has line symmetry. Ask: How can we look at individual designs and test them for line symmetry? By drawing horizontal and vertical lines across and and down through the center of individual designs, we can determine if they have line symmetry. For example, "snails" has both horizontal and vertical line symmetry; "tiger eyebrow" has vertical line symmetry; and "heart" has neither horizontal nor vertical line symmetry.

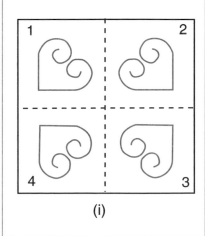

(i)

Fig. 2.3. Fold-and-trace method

Fig. 2.4. Completed paj ntaub

6. Ask students to write their own definitions of *line* and *rotational symmetry*. Share within groups and the whole class. Have the class come to consensus on these definitions.

7. To complete the paj ntaub, find the center of each square by determining the intersection of the square's vertical and horizontal axes. Glue each square to the next larger in size, so that the centers of successive squares are touching. As a final step, cut a small triangle from the cardboard, to be used as a stencil. Using this stencil, cut additional triangles out of construction paper as a border around the largest square (see fig. 2.4).

8. The paj ntaub lends itself to other mathematical ideas, such as area and similarity. For example, after the $8'' \times 8''$ square is attached to the $6'' \times 6''$ square, ask: How can we find the area of the border between the $6'' \times 6''$ square and the $8'' \times 8''$ square? Ask students to determine this, working in groups. Examine for the $10'' \times 10''$ square as it is attached to the other squares.

9. For another extension, ask students to record the length, width, perimeter, and area of each square. Find ratios to compare lengths of sides, widths of sides, perimeters, and areas of pairs of squares. What conclusions can be drawn? What would happen to the perimeter and area if the dimensions of the $6'' \times 6''$ square were doubled or tripled? Would we obtain similar conclusions for rectangles, triangles, or circles?

These activities, used by themselves, provide only a cultural add-on. However, when the activities are provided in context, by including the history of the Hmong people and the significance of the paj ntaub to their culture, the activities become much more meaningful to students. Including the mathematics activities as part of an interdisciplinary theme further strengthens their impact, and the topic lends itself to interdisciplinary connections. As examples, the paj ntaub mathematical activities can become part of units on Southeast Asia, the Vietnam War, or immigration. Whichever theme you choose, make the unit multidisciplinary by including Hmong songs, games, folktales, and children's literature. If you live in an area populated by Hmong refugees, invite a speaker to visit your class or take a field trip to a Hmong cultural event. *Hmong Folklife* by Don Willcox (1986) is written for adults but contains interesting information about Hmong life and culture. *The Whispering Cloth*, by Pegi Deitz Shea (1995), is a children's book that tells the story of a child in a refugee camp learning to embroider the paj ntaub. It is especially appropriate for use with the paj ntaub activities.

An important aspect of studying other cultures is discovering that different peoples share common experiences and customs. We have included some examples of cross-cultural connections that would be appropriate for the classroom.

Hmong and Navajo Connections

The Hmong changed their embroidery significantly to attract western customers. The same is true of Navajo weavers. Originally, the Navajo wove blankets. Traders persuaded them to weave heavier rugs and to place borders around the rugs. Navajo weavers always leave a small part of the border, the *spirit line*, unfinished so that the spirit of the weaver will not be trapped inside the rug. Similarly, Hmong women always embroider one section of the border with a different color thread. *The Magic of Spider Woman*, by Lois Duncan (1996) is a children's book that retells the origins of Navajo weaving.

Hmong and Latino Connections

Hmong women began to embroider story cloths to record their former life. In another part of the world, central Mexico, the women of Lake Patzcauro embroider pictures of their former life. The children's book, *Life around the Lake*, by Mariel E. Presilla and Gloria Soto (1996), tells the story of these women and includes beautiful illustrations that are hauntingly similar to Hmong story cloths.

Hmong and American History Connections

The history of American pioneers also provide connections to the Hmong experience. Like the Hmong, many pioneers came to America to escape persecution and find a better life. Quilts were created by pioneer women from scraps of material. They were created in individual squares with symmetric designs. Each design had its own name. Two children's books about quilts and pioneer life are *The Quilt-Block History of Pioneer Days* by Mary Cobb (1995) and *Selina and the Bear Paw Quilt* by Barbara Smucker (1995).

Finally, encourage students in your class to find out more about their own personal history. Many students will have families who, at one time, may have faced great difficulty and possibly danger. Wherever their ancestors emigrated from, they too were once immigrants in our country. Ask students to bring in crafts, stories, and other mementos that are significant to their own culture. Invite parents and grandparents to share cultural information.

CONCLUSION

The examples in this article illustrate that mathematics is everywhere, is used by every culture, and can easily be integrated with other subjects. The study of a cultures' textiles often reveals a great deal about the history and customs of the culture. All cultures produce textiles, and many cultures use similar methods to produce their textiles and create shapes that are very much alike. The study of one particular culture can encourage students to be more accepting of cultural differences. The mathematics activities suggested here represent just a few ideas about how to use culture to bridge differences between people and to make mathematics accessible to all.

THANKS

Many thanks go to Sue Carey, art teacher in the City of Eau Claire School System, Eau Claire, Wisconsin, for contributing the draft graphics for this article.

REFERENCES

Chan, Anthony. *Hmong Textile Designs.* Owings Mill, Md.: Stemmer House, 1990.

Cobb, Mary. *The Quilt-Block History of Pioneer Days.* Brookfield, Conn.: The Millbrook Press, 1995.

Duncan, Lois. *The Magic of Spider Woman.* New York: Scholastic, 1996.

National Council of Teachers of Mathematics. *Professional Standards for Teaching Mathematics.* Reston, Va.: National Council of Teachers of Mathematics, 1991.

Presilla, Maricel E., and Gloria Soto. *Life around the Lake: Embroideries by the Women of Lake Patzcuaro.* New York: Henry Holt and Company, 1996.

Roop, Peter, and Connie Roop. *The Hmong in America: We Sought Refuge Here.* Appleton, Wis.: League of Women Voters, 1990.

Shea, Pegi Deitz. *The Whispering Cloth.* Honesdale, Pa.: Boyds Mill Press, 1995.

Sleeter, Christine E. "Mathematics, Multicultural Education, and Professional Development." *Journal for Research in Mathematics Education* 28 (1997): 680–96.

Smucker, Barbara. *Selina and the Bear Paw Quilt.* New York: Random House, 1995.

Willcox, Don. *Hmong Folklife.* Penland, N.C.: Hmong Natural Association of North Carolina, 1986.

Using Asian-Pacific Literature to Enhance Problem Solving in the Elementary School Mathematics Classroom 3

Nancy C. Whitman

I don't understand the problem." "I don't know what to do." These are common comments by students who have difficulty solving verbal problems in mathematics. A fairly common practice is for the teacher to rephrase the problem until the student understands it and says "Is that what they wanted?" or "Is that all?" It is likely that students who do not understand how to solve a given problem may not comprehend the context and setting of the problem. Using a story or a piece of children's literature as background provides the students with a setting and context for the problem that they can all understand. With a story they now share a common experience. No student is disadvantaged because he or she does not have the necessary experience for understanding the problem situation.

Stories are valuable in informing us about how mathematical ideas are developed (Whitman 1979) and in providing motivation and interest. Stories also allow the teacher to integrate her or his teaching of mathematics with other areas of study, such as language arts and social studies. By doing this, the teacher is assisting the students in seeing and making connections from mathematics to other subject areas.

By integrating mathematics with other areas of study, the teacher addresses the continuing concerns of educators about juggling the curriculum to include all disciplines. Hence, integration may also serve as a partial managerial solution for the classroom teacher. That integration with other fields of study can provide for meaningful mathematics learning is reflected in the professional literature (Bertheau 1994; Karp 1994; Whitman 1979; Wong 1991). One of the standards for the mathematics curriculum, according to the National Council of Teachers of Mathematics (NCTM), is that students make mathematics connections with other fields of study (NCTM 1989).

Using funds from an Eisenhower grant, a workshop for teachers was conducted to help them see and use the links between mathematics and literature. The focus was on Asian-Pacific literature because many of the classroom students were of that ethnic background. It was believed that lessons that related to the students' cultural background were needed since it was important for the students to experience relationships, characters, and settings with which they could identify. This also would afford an opportunity for students to share and learn more about their own culture and the cultural backgrounds of their classmates.

After the teachers in the workshop were shown examples of the successful integration of mathematics problem solving and literature in the classroom, they reviewed over a hundred children's books and selected a few to try out in their own classes. They created, implemented, and evaluated lessons they felt were appropriate for their students. Many of the lessons incorporated the teaching of English and mathematics. What follows is one of the many successful lessons used in the classroom, along with its evaluation by the teacher.

SAMPLE LESSON

Grade Level: Grade Two

Book Used: *The Magic Amber: A Korean Legend*
                by Charles Reasoner

Story Summary: An old rice farmer and his wife are kind and generous. One day they are repaid for their kindness and generosity with the gift of a magic amber that supplies them endlessly with a pot of rice. Lon Po hears of this and disguises himself as a beggar and steals the magic amber. However, he loses it in the river. Although the rice farmer and his wife no longer had rice to give away, their neighbors continued to bring them gifts. One day a catfish is given to the rice farmers. Inside this fish is the magic amber.

Materials

Computer, construction paper, and crayons.

Time

One to three class periods of 50 minutes each. (Some students took longer to create and illustrate their problems.)

Objectives

1. To have students write one word problem that relates to the story and show work to solve it
2. To work at solving each other's problems.

Procedure

1. Discuss the use of the word *amber.* Relate it to past uses and to its use in the title of the story.
2. Read the story.
3. Discuss the main character traits of Lon Po and the old man.
4. Have the students predict events as the story is being read.
5. Discuss the characters in the story so that the students will know what characters they can include in their word problems.
6. Review past creations of word problems from stories read. In particular, discuss the need to raise questions based on the information provided.
7. Have the students create problems of their own and type them on the computer.
8. Have the students illustrate their problems and show how to solve them.
9. Consolidate the problems that were placed on computer.
10. Have the students solve the list of consolidated problems. *Note:* If the combined problems do not have illustrations, encourage the students to draw their own.

The classroom procedures provided for the integration of mathematics, English, and computer usage. It afforded an opportunity for students to create drawings to help them solve problems. The classroom activities gave the teacher situations in which to talk with the students about the proper usage of drawings to enhance the solving of problems. The story provided a common setting and context for all the students.

EXAMPLES OF PROBLEMS CREATED, ILLUSTRATED, AND SOLVED BY STUDENTS

The problems below are the verbatim problems of the students. They are based on the students' imagination, creativity, past experiences, and exposure to writing story problems. In some cases, the pictures drawn by the students did not depict the situation expressed in the problem. This afforded the teacher an opportunity to discuss the purpose of pictures in solving verbal problems and to encourage a more accurate portrayal of the problem in the drawings.

1. There was one bag of rice that the old man needed. Lon Po wanted three in return. If the old man wanted four bags of rice, how many would Lon Po want in return?

   Solution:

old man	1	2	3	4
Lon Po	3	6	9	12

   No drawing accompanied this problem. Instead the table above was provided.

2. There are 16 bags of rice and 17 people. Can each person have a bag of rice?

   Solution:

   No. A one-to-one matching was drawn of people to bags of rice. One person was not matched to a bag of rice. (See fig. 3.1.)

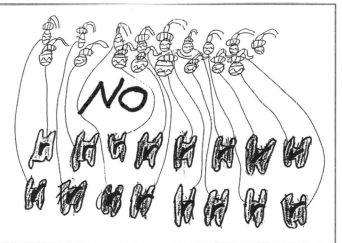

There are 16 bags of rice and 17 people. Can each person have a bag of rice?

Fig. 3.1. Seth Nelson's word problem

3. The old man put 214 pieces of rice in a bag. The old woman put 117 pieces of rice in the bag. How many pieces of rice are in the bag altogether?

   Solution: 214
   　　　　　　　+117
   　　　　　　　331

   A picture was drawn of the old man and the old woman placing rice in separate bags. (See fig. 3.2.)

The old man put 214 pieces of rice in a bag. The old woman put 117 pieces of rice in the bag. How many pieces of rice are in the bag altogether?

Fig. 3.2. Jan's word problem

Lon Po falls off a boat. He loses the Amber in the river. So many fishes saw it but only one ate it. How many fishes did not eat the Amber? Loop your answer.

(a) 1 fish

(b) 2 fishes

(c) not enough information

Fig. 3.3. Michael's word problem

The old man has 9 bags of rice. He gave 3 bags to the neigh-bors. How many are left?

Fig. 3.5. DeLarian's word problem

4. Lon Po falls off a boat. He loses the Amber [*sic*] in the river. So many fish-es [*sic*] saw it but only one ate it. How many fishes [*sic*] did not eat the Amber [*sic*]? Loop your answer. (See fig. 3.3.)

   (a) 1 fish

   (b) 2 fishes [*sic*]

   (c) not enough information

5. There are 10 fishermen. Each fisherman caught 10 fishes [sic] How many did they catch in all?

   Solution: $10 \times 10 = 100$

   A picture of fishermen catching fish was drawn. It did not show that each had caught 10 fish. This provided a situation for the teacher to discuss the need for a picture to reflect the circumstances of the problem. (See fig. 3.4.)

There are 10 fishermen. Each fisherman caught 10 fishes. How many did they catch in all?

Fig. 3.4. Christopher's word problem

6. The old man has 9 bags of rice. He gave 3 bags to the neighbors. How many are left?

   Solution:
   $$\begin{array}{r} 9 \\ -3 \\ \hline 6 \end{array}$$

   A picture was drawn of the old man going to the neighbors with a bag of rice in his hands. The drawing did not display the old man giving three bags of rice away, nor did it show he had nine bags initially. Again, this provided opportunity for discussing the need for an accurate drawing of circumstances for problem-solving purposes. (See fig. 3.5.)

Teacher Evaluation

The teacher commented that the students enjoyed the story, and they enjoyed creating their own word problems and solving those of others. The teacher was amazed at the different types of word problems the students created; some created more than one problem. The teacher also found this type of lesson useful for assessment purposes. She noted that some of the students had difficulty in wording their problems and needed help in using words as "altogether" and "in all."

Note: This classroom was a "whole-language" classroom; therefore, mathematics was integrated with language teaching as much as possible.

At the end of the workshop, the teachers were asked to select about three books they would like to have for their future use based on the books they had reviewed and used in their classes or those some other teachers had reviewed and used. What follows are the teachers' choices with annotations and suggestions on topics of mathematics with which they may be used.

Au, Joy S. Illustrated by Jill Chen Loui. *Going to the Tide Pools in Hawaii Nei*. Honolulu: MnM Books, 1995. ISBN 1-56647-082-X

Children discover twelve sea creatures in the tide pools in Hawaii. (Classification)

Chinn, Karen. Illustrated by Cornelius Van Wright and Ying-Hwa-Hu. *Sam and the Lucky Money*. New York: Lee and Low Books, 1995. ISBN 1-880000-13-X

Sam must decide how to spend the lucky money he has received for Chinese New Year. (Money activities)

Demi. *The Empty Pot*. New York: Henry Holt, 1990. ISBN 0-8050-1217-6

When Ping admits that he is the only child in China unable to grow a flower from the seeds distributed by the emperor, he is rewarded for his honesty. (Seasons of the year)

Flack, Majorie, and Kurt Wiese. *The Story about Ping*. New York: Penguin Group, 1966. ISBN 0-14-050-2416

A little duck finds adventure on the Yangtze River when he is too late to board his master's houseboat one evening. (One-digit and two-digit addition and sequencing)

Friedman, Ina R. Illustrated by Allen Say. *Chopsticks in America*. New York: Houghton, Mifflin, 1984. ISBN 0-395-35379-3

An American sailor courts a Japanese girl and each tries, in secret, to learn the other's way of eating.

Guback, Georgia. *Luka's Quilt*. New York: Greenwillow, 1994. ISBN 0-688-12154-3

Luka and her grandmother disagree over the colors a traditional Hawaiian quilt should include when her grandmother makes her a quilt. (Line and rotational symmetry)

Hillman, Elizabeth. Illustrated by John Wallner. *Min-Yo and the Moon Dragon*. New York: Harcourt Brace Jovanovich, 1992. ISBN 0-15-254230-2

When the moon suddenly appears to be approaching the earth, Min-Yo climbs the cobweb staircase between the earth and moon to ask the moon dragon for help. (Distance, weight)

Ho, Minfong, and Saphan Ros. Illustrated by Jean and Mou-Sien Tseng. *The Two Brothers*. New York: Lothrop, Lee, & Shepard Books, 1995. ISBN 0-688-12550-6

Brought up in a Buddhist monastery, two brothers go out into the world to very different fates armed with the advice of a wise abbot.

Hulme, Joy N. Illustrated by Carol Schwartz. *Sea Squares*. New York: Hyperion Books for Children, 1991. ISBN 1-56282-520-8

Rhyming text and illustrations of such sea animals as whales, gulls, clown fish, and seals are provided. (Counting and squaring numbers from one to ten)

Laurin, Anne. Illustrated by Charlkolaycak. *Perfect Crane*. New York: HarperCollins, 1981. ISBN 0-06-023743-0

A lonely Japanese magician gains friends through the paper crane that he brings to life but then must set free. (Seasons of the year)

Louie, Ai-Ling. Illustrated by Ed Young. *Yeh Sheng: A Cinderella Story from China*. New York: Philomel Books, 1982. ISBN 0-399-61203-3

A young Chinese girl overcomes the wickedness of her stepsisters and stepmother to become the bride of a prince.

Nunes, Susan. Illustrated by Cissy Gray. *To Find the Way*. Honolulu: University of Hawaii Press, 1992. ISBN 0-8248-1376-6

Using his knowledge of the sea and stars, Vahi-roa the navigator guides a group of Tahitians abroad a great canoe to the unknown islands of Hawaii.

Rattigan, Jama Kim. Illustrated by Lillian Hsu-Flanders. *Dumpling Soup.* Boston: Little, Brown, and Co., 1993. ISBN 0-316-73445-4

A young Korean girl in Hawaii tries to make dumplings for her family's New Year celebration. (Shopping for ingredients, counting backwards, ordinal numbers)

Tomiko, Chiyoko. Illustrated by Yoshiharu Tsuchida. *Rise and Shine, Mariko-chan.* New York: Scholastic, 1986. ISBN 0-590-45507-9

The youngest of three sisters in a Japanese family gets ready for a day in preschool. (Time, sequence)

Tompert, Ann. Illustrated by Robert Andrew Parker. *Grandfather Tang's Story: A Tale Told with Tangrams.* New York: Crown Publishers, 1990. ISBN 0-517-57487-X

(Area, tangrams, transformations)

Williams, Jay. Illustrated by Mercer Mayer. *Everyone Knows What a Dragon Looks Like.* New York: Macmillan,1976. ISBN 0-02-045600-X

Because of the road sweeper's belief in him, a dragon saves the city of Wu from the Wild Horsemen of the north. (Good for discussing sequencing—1st, 2nd, 3rd)

Yep, Lawrence. Illustrated by Jean and Mou-sien Tseng. *The Boy Who Swallowed Snakes.* New York: Scholastic, 1994. ISBN 0-590-46168-0

A brave and honest young boy with an unusual appetite for snakes turns the tables on a greedy rich man. (Rate of increase of snakes if doubled, tripled, etc.; study of patterns)

Young, Ed. *Lon Po Po: A Red-Riding Hood Story from China.* New York: Philomel Books, 1989. ISBN 0-399-21619-7

Three sisters staying home alone are endangered by a hungry wolf who is disguised as their grandmother.

————. *Cat and Rat: The Legend of the Chinese Zodiac.* New York: Henry Holt, 1995. ISBN 0-8050-2977-X

According to Chinese legend, the twelve animals of the zodiac were selected by the Jade Emperor after he invited all the animals to participate in a race. This is the story of that race. It is also the story of Cat and Rat and why they will always be enemies. (Sequencing numbers and learning about calendars)

Lon Po Po: A Red-Riding Hood Story from China was chosen by four teachers; *Yeh Sheng: A Cinderella Story from China* by three teachers; and *Dumpling Soup, Cat and Rat: The Legend of the Chinese Zodiac* and *Empty Pot* were chosen by two teachers.

REFERENCES

Bertheau, Myrna. "The Most Important Thing Is...." *Teaching Children Mathematics* 1 (October 1994): 112–5.

Karp, Karen S. "Telling Tales: Creating Graphs Using Multicultural Literature." *Teaching Children Mathematics.* 1 (October 1994): 87–91.

National Council of Teachers of Mathematics. *Curriculum and Evaluation Standards for School Mathematics.* Reston, Va: National Council of Teachers of Mathematics, 1989.

Whitman, Nancy C. "Mauka-Makai and Other Directions." In *Applications in School Mathematics*, edited by Sidney Sharron and Robert E. Reys, pp. 59–69. 1979 Yearbook of the National Council of Teachers of Mathematics. Reston, Va: National Council of Teachers of Mathematics, 1989

Wong, Raylene. *A Kindergarten Multicultural Literature-Based Problem-Solving Curriculum Plan.* Masters Paper. Department of Curriculum and Instruction, University of Hawaii at Manoa, 1991.

Differences in Arithmetic Nomenclature

4

Jean Bee Chan

In this article, I will discuss two differences in verbal arithmetic in Chinese and in English. Of course, written arithmetic is the same, using the decimal numeral system in Arabic notation.

As an immigrant from China, I entered college in the United States after my K–12 education in Hong Kong and mainland China. Besides the usual lack of adequate English skills, I encountered culture shock in verbal arithmetic in two areas.

First, in English, fractions are verbalized by naming the numerators first, followed by the denominators. In Chinese, the practice is the opposite. For example, the fraction 2/3 is expressed "two-thirds" in English but "three parts two" in Chinese. In Chinese, "three parts two" means the whole is divided into three equal parts and two parts of the three are represented in the fraction.

The addition 2/3 + 2/5 = 10/15 + 6/15 = 16/15 = 1 1/15 is verbalized as "three parts two" plus "five parts two" equals "fifteen parts ten" plus "fifteen parts six." The sum equals "fifteen parts sixteen," which is "fifteen parts fifteen" plus "fifteen parts one" or "one" and "fifteen parts one." As school children, we verbalized adding and subtracting fractions and recited the multiplication table for "fun activities" in class.

Another major verbal difference is scaling large numbers. Large numbers are expressed in terms of *won* (in Mandarin, the official spoken Chinese), meaning 10 000. So, a hundred thousand is 10 won, a million is 100 won, and so on. This small change of one zero in scaling continues to be confusing to me, even after teaching mathematics in English for twenty-five years.

Now, I must write down numbers I hear in English to make sure I understand them correctly. I must quickly do calculations mentally in Chinese and slowly translate the results to English. It is difficult for me to verbalize English calculation rapidly to this day because my multiplication table is still in Chinese.

Koreans and Japanese say fractions and scale large numbers the same way Chinese do. Students from those countries may also have similar problems verbalizing arithmetic.

Counting in Chinese, Japanese, and Korean
Support for Number Understanding **5**

Irene T. Miura
Yukari Okamoto

The recent Third International Math and Science Study (TIMSS 1996) examining cross-national differences in mathematics achievement has once again underscored the superior performance of students from Asian language–speaking countries, particularly Singapore, Korea, Japan, and Hong Kong. Teaching strategies and parental support for education are being cited as important factors in explaining these differences. However, without significant societal change in parental attitudes toward, and involvement in, mathematics education or a shift in our understanding of the meaning of individual student motivation that provides the educational context for culturally mediated teaching methods, our ability to learn from these comparisons and apply that understanding to the U.S. classroom setting is severely limited.

The language of mathematics is missing from the current discussion of cross-national achievement comparisons (c.f., California Education Policy Seminar [1997]), even though there is a renewed interest among researchers in the role that cultural processes may play in the understanding of mathematics and in the performance of mathematics tasks (c.f., Cobb [1993]; Kaput [1991]; Saxe [1988]). Much of this research is influenced by the work of Soviet psychologist L. S. Vygotsky and that of his followers (c.f., Luriia [1976]; Vygotsky [1962, 1978]). Central to these processes is the child's acquisition of cultural tools for thinking and learning—one of the most important of which is language (Rogoff 1990; Steffe, Cobb, and von Glasersfeld 1988). Mathematical activities cannot be separated from the cultural contexts in which they are set, and this is particularly true for the language of mathematics. As Kaput (1991, p. 55) puts it:

> One can draw an analogy between the way the architecture of a building organizes our experience, particularly our physical experience, and the way the architecture of our mathematical notation system organizes our mathematical experience. As the physical level architecture constrains and supports our action in ways that we are often unaware of, so do mathematical notation systems.

Nunes (1992) has interpreted Vygotsky's writings to suggest that the use of culturally developed symbol systems restructures mental activity without altering basic abilities such as memory and logical reasoning. As an example, Nunes cites numeration systems (like the English-language base-ten system) that allow humans to go beyond their natural memory capacities to count large numbers of objects. By contrast, the counting system of the Oksapmin of Papua New Guinea uses body parts; in this method they are able to count up to twenty-four in some cases and sixty-eight in others (Saxe 1982). This means, of course, that when counting a large number of objects, someone counting in English will be able to outperform someone counting in Oksapmin because he or she is using a better counting tool. It is not that the English speaker's memory for digits is necessarily greater than that of the Oksapmin; it is just that the structure of the counting tool is more efficient for counting, thereby making the activity easier to perform.

This notion that reasoning ability, per se, does not differ across cultures is also supported by the work of Okamoto, Case, Bleiker, and Henderson (1996) testing Case's developmental theory (1992) using novel, non-school-related tasks. In a study of Japanese and U.S. six-, eight-, and ten-year-olds, they found no differences between children in the two countries in their levels of intuitive understanding of number and quantitative reasoning with a balance beam, with one exception: number understanding at age six. Japanese six-year-olds showed an eight-year-old level of number understanding. However, by age eight, the U.S. children's performance on these tasks equaled that of the Japanese.

Miller and Stigler's work (1987) comparing Chinese and U.S. children's counting performance also supports Nunes's hypothesis. Although the Chinese children in their study counted significantly better than did the U.S. children, with respect to control of counting principles (Gelman and Gallistel 1978), the children were quite similar. Thus, it appears that the counting performance (but not the counting understanding or use of counting to solve problems) of the Chinese children may be supported by the numeration system. However, counting provides a foundation for acquisition of number concepts, and it is our view that certain characteristics of Asian number languages (culturally developed tools for mathematics) may promote a developmental head start and affect the later performance of mathematical tasks.

EFFECTS OF AN ASIAN NUMBER-NAMING SYSTEM ON MATHEMATICS TASKS

Throughout this chapter, we use the term *Asian language* to refer to languages that are rooted in ancient Chinese (among them, Chinese, Japanese, and Korean). Although, for ease of reading, our examples usually describe an effect in only one of these languages, the information can be generalized to all three. The counting or number-naming systems in these languages are organized so that they are congruent with the traditional base-ten numeration system. In this system, the value of a given digit in a multidigit numeral depends on the face value of the digit (0 through 9) and on its position in the numeral, with the value of each position increasing by powers of ten from right to left. The spoken numerals in western languages (e.g., *twelve* and *twenty* in English and *four-twenty* for 80 in French) may lack the elements of tens and ones that are contained in them. Also, the order of spoken and written numerals may not agree (e.g., fourteen for 14 in English and three-and-forty for 43 in German). In these Asian languages, ten is treated as a discrete unit so that the numeral 11 is read as ten-one, 12 as ten-two, 14 as ten-four, 20 as two-ten(s), 43 as four-ten(s)-three, and 80 as eight-ten(s). Fourteen and 40, which are phonologically similar in English, are differentiated in Japanese; 14 is spoken as ten-four and 40 as four-ten(s). Plurals are tacitly understood; thus, the spoken numeral corresponds exactly to the implied quantity represented in the written form (table 5.1). Furthermore, when character number symbols are used in place of the Arabic numerals, the correspondence between spoken and written numerals is even more precise; for example, the numeral 46 is written in character symbols as four-ten(s)-six.

Effects on Number Representation

The Japanese spoken numeral describes precisely what is represented by the base-ten numeration system. The syntax and semantics are, as Cobb (1993) suggests, reflexively related. Place value (i.e., the meaning of tens and ones in a two-digit numeral) also appears to be an inherent element of those representations (Miura et al. 1994). For example, when asked to show or construct the numeral 42 using base-ten blocks (unit blocks and bars that have ten segments

Table 5.1
Number Names in Four Languages

Number	English	Chinese	Japanese	Korean
1	one	yi	ichi	il
2	two	er	ni	ee
3	three	san	san	sam
4	four	si	shi	sah
5	five	wu	go	oh
6	six	liu	roku	yook
7	seven	qi	shichi	chil
8	eight	ba	hachi	pal
9	nine	jiu	kyu	goo
10	ten	shi	juu	shib
11	eleven	shi-yi	juu-ichi	shib-il
12	twelve	shi-er	juu-ni	shib-ee
13	thirteen	shi-san	juu-san	shib-sam
14	fourteen	shi-si	juu-shi	shib-sah
15	fifteen	shi-wu	juu-go	shib-oh
16	sixteen	shi-liu	juu-roku	shib-yook
17	seventeen	shi-qi	juu-shichi	shib-chil
18	eighteen	shi-ba	juu-hachi	shib-pal
19	nineteen	shi-jiu	juu-kyu	shib-goo
20	twenty	er-shi	ni-juu	ee-shib
21	twenty-one	er-shi-yi	ni-juu-ichi	ee-shib-il
22	twenty-two	er-shi-er	ni-juu-ni	ee-shib-ee
23	twenty-three	er-shi-san	ni-juu-san	ee-shib-sam
30	thirty	san-shi	san-juu	sam-shib
40	forty	si-shi	shi-juu	sah-shib
50	fifty	wu-shi	go-juu	oh-shib
60	sixty	liu-shi	roku-juu	yook-shib
70	seventy	qi-shi	shichi-juu	chil-shib
80	eighty	ba-shi	hachi-juu	pal-shib
90	ninety	jiu-shi	kyu-juu	goo-shib

marked), American, Swedish, and French first-grade children were more likely than their Asian counterparts initially to represent the numeral using a one-to-one collection (i.e., 42 unit blocks to show 42). Chinese, Japanese, and Korean first graders represented the numeral with a canonical base-ten representation (i.e., 4 tens blocks and 2 unit blocks for 42) (table 5.2). The Asian-language speakers differed significantly from the non-Asian-language speakers in the kinds of constructions they made. For the five numerals presented in the study (11, 13, 28, 30, and 42), the Asian-language speakers used more canonical base-ten representations (e.g., 2 tens blocks and 8 unit blocks for 28) than did the non-Asian language speakers. The Asian-language speakers used fewer one-to-one collection representations than did the others (Miura et al. 1994).

Table 5.2
Means for Cognitive Representation of Number Category by Language Groups

Category	Asian Languages			Non-Asian Languages		
	Chinese	Japanese	Korean	United States	France	Sweden
Canonical base-ten	4.04	3.75	4.83	0.38	0.39	0.57
One-to-one collection	0.48	0.45	0.04	4.13	3.96	4.44

Note: The total possible was 5.

Effects on Counting

Studies comparing Chinese- and English-speaking children's counting behavior (Miller and Stigler 1987; Miller et al. 1995) showed that Chinese-speaking children exhibit the same pattern of development of counting skills as do American preschoolers. However, the Chinese children made significantly fewer errors in number naming than did U.S. children. The activity of generating number labels poses little problem for Chinese children, and in this aspect of counting, they surpass their American counterparts. The functional organization of counting differs across the two systems and this appears to give Chinese children an advantage that is clear when the skill is taken as a whole.

Counting from 1 to 9 in an Asian-language system is automatized; one counts from 1 to 10, and then the 1 to 9 is reinforced in the numerals 11 through 19. As a result, mathematics tasks around tens are easier because the child deals with 1 to 9 only. Learning the teen number names is particularly troubling for U.S. children (Miller et al. 1995). In addition, Chinese-speaking children do not have to learn the decade number names, a requirement which often impedes U.S. children's counting performance.

Effects on Place-Value Understanding

On a set of five place-value tasks, Japanese (Mean = 2.96) and Korean (Mean = 4.42) first graders showed significantly greater understanding of place-value concepts than did children from the United States (Mean = 1.46), France (Mean = 0.74), and Sweden (Mean = 0.70) (Miura et al. 1993). All of the Japanese and Korean first graders completed at least one problem correctly; 42 percent of Japanese children and 54 percent of the Korean first graders were able to solve all five problems correctly. By contrast, the Swedish, French, and U.S. children in the study performed significantly less well; 50 percent of the U.S. first graders could not solve any of the problems correctly. The children had not been taught place-value concepts in school prior to the testing. It was our conclusion that the children's performance was related to the structure of their respective number languages. Number-language systems that supported (or did not support) the place-value concepts inherent in the intuitive understanding of what a particular numeral might mean were thought to account for the differential understanding of place value exhibited by the children.

Effects on Addition and Subtraction Performance

In the specific area of addition and subtraction with regrouping (borrowing/carrying) that is taught at the second-grade level in most countries, the Asian number-language system may have an important effect. In the addition algorithm, $59 + 8 = ?$ (table 5.3a), the numbers in the ones column total 17. English speakers must ask themselves if they can regroup or trade to make a ten. If the answer is yes, this results in a regrouped 1 ten and 7 ones. The 7 is written in the ones column, and the 1 (which represents 1 ten) is added to the 5 in the tens column (table 5.3b). In Japanese, the sum of the numbers in the ones column is spoken as ten-seven. Therefore, it is understood that a ten is formed, and 1 ten is carried to the tens column, eliminating the intermediate assessment step and keeping the meaning of the individual digits in the solution intact (table 5.3c).

Consider the following subtraction problem, $67 - 59 = ?$ (table 5.4a). There are at least two possible solutions for this problem after determining that there are not enough ones to solve the problem without regrouping. First, 1 ten from the tens column can be traded or regrouped for 10 ones to make the number in the ones column a 17. Then, the problem is solved by subtracting 9 from 17 to equal

Table 5.3
Addition with Regrouping

	a	b	c
		1	10
	59	59	59
	+ 8	+ 8	+ 8
		7	7

8 (table 5.4b). If the regrouped 17 is spoken as ten-seven, a second solution becomes possible: 9 can be subtracted from the 10 to leave a remainder of 1, which is then added to the remaining 7 to equal 8 in the ones column (table 5. 4c).

U.S. children are taught to subtract using the trading or regrouping algorithm (table 5.4b) (e.g., Addison-Wesley [1993]). The terms vary according to the text-book publisher. This first solution is used by Asian children as well. The second solution (table 5.4c) can also be used easily by Asian children because their number language treats the teen numbers (11–19) as 10 + the single digit; the two numbers (10 and the single digit) can be dealt with separately in the solution process. While English-speaking children must learn the addition and subtraction facts to eighteen, Asian-language-speaking children do not have to master combinations beyond ten.

Table 5.4
Subtraction with Regrouping

	a	b		c	
		5 1		5 10	
	67	67	17 − 9 = 8	67	10
	− 59	− 59		− 59	− 9
		8			1 + 7 − 8

Mathematics involves the manipulation of symbols, especially numerals, that must be understood before they can be used efficiently for problem solving. Children may be hindered in their ability to perform certain mathematical tasks, such as addition and subtraction with regrouping, unless they understand the meaning of the digits in a multidigit numeral. For example, in order for U.S. children to use the 10 + the single digit method in regrouping, they must first be taught that the teen numbers are, indeed, composed of ten and a single digit (Fuson, Fraivillig, and Burghardt 1992).

Counting appears to be one method by which children may learn place-value concepts (Miura et al. 1993). Asian language speakers have a distinct advantage in learning these concepts because their languages treat ten as a discrete unit. Thus, for English-language speakers, whose number-naming system lacks this treatment of ten, other teaching strategies must be employed (c.f., Bell [1993]; Freitag [1993]). The following exercises are for grades K–3 to assist children to develop an understanding of the base-ten foundation of numbers. Most of these exercises are not new; they have already been introduced and promoted through teacher in-service and conference workshops. What is new is the research support for why these strategies may be influential in developing

COUNTING IN THE CLASSROOM: ACTIVITIES FOR LEARNING ABOUT NUMBERS

place-value understanding and promoting subsequent mathematical problem solving. Furthermore, rather than being used simply as introductory tools or for enrichment purposes, it is recommended that these exercises be continued throughout the K–3 years to reinforce these base-ten concepts.

Learning to Count with 10 as a Discrete Symbol

Using the typical counting book, preschool and kindergarten children learn to count by looking at a numeral, saying the number name, and pointing to the correct number of objects in one-to-one correspondence. When the numeral is 11, the child continues to enumerate 11 objects. The concept that 11 is 10 and 1 more is lacking in this exercise and must be taught later. Instead, teachers are encouraged to use items that result in 10 being a separate object made up of individual units: e.g., pieces of pizza resulting in 10 pieces making a whole; petals on a flower; flowers on a stalk; or individual pieces resulting in a geometric whole. The teen numbers, by this method, would be depicted as 10 + the individual digit, rather than collections of individual items.

More Counting with 10 as a Unique Symbol

Baroody (1987) described an experiment that taught first graders a notation system in which 1 was a straight line (an I) and 10 was a unique symbol (an upside-down U). He called the system *Egyptian numbers*. Using this notation system, first graders were successfully taught addition and subtraction with regrouping. The children eventually progressed from Egyptian numbers to standard notation. When they encountered difficulty, they were reminded to think of the algorithm in Egyptian numbers, a strategy that helped them to visualize the problem, overcome the difficulty, and solve it.

In an experiment in the laboratory preschool/kindergarten at San Jose State University, children were taught to count in a variety of bases by using "T" as the spoken symbol for the base (R. Spaulding, personal communication, April 1987). For base ten, the system is as follows: 1, 2, 3, ..., 9, T, T-1 (11), T-2 (12), T-3 (13), ...,T-9 (19), 2-T (20), 2-T-1 (21), 2-T-2 (22), etc. After thoroughly understanding the system, children were told that another name for T-1 is eleven; another name for T-2 is twelve, etc. Phonologically, many of the decades in base ten (e.g., 40, 60, 70, 80, and 90) sound similar to the numerals as we typically recite them. According to Spaulding, children could count in any base (1–10) using this system. It allowed for a deep understanding of the structure of the number-naming system, especially the teen numbers that are generally problematic for children learning to count.

Both of these novel counting systems treat 10 as a unique symbol, and this may benefit children as they learn number concepts up through 99. These counting systems may be especially useful in helping children to visualize the teen numbers and thereby promote understanding.

Counting in an Asian System

Introduce numbers from an Asian language when children have mastered the English number names from 1 through 99, including the number that comes before x and the number that comes after x (e.g., What is the number before 15? What is the number after 15?). Teach the number names from 1 to 10 in Chinese, Japanese, or Korean (both verbal and written [romanized] forms) and have children write out the number words next to the corresponding numerals from 1 through 99. The simplicity of the number generating system, as well as the meaning of the individual digits in the multidigit numerals, becomes apparent through this exercise. Additional mathematics teaching materials using a variety of Asian languages have been developed by and are available from the Northern Rivers Mathematical Association in Australia (Bell 1993).

Using Manipulative Materials

A common activity is a count and record exercise in which the child uses an 8-1/2" by 11" piece of paper (a mat) divided into three columns and marked at the top (from right to left) with the words *ones*, *tens*, and *hundreds*. As each item is placed in the ones column, the child writes the numeral on a recording sheet. When there are nine items in the ones column, the next numeral on the mat is represented by a discrete unit representing 1 ten. Examples of items that can be used for this exercise are (*a*) beans, bean sticks (ten beans glued by the child on to an ice cream stick), and bean rafts (ten bean sticks glued onto an ice cream stick frame); (*b*) commercially available base-ten blocks; and (*c*) pennies, dimes, and dollars (from a math textbook kit).

Encouraging the Maintenance of an Asian First Language

Children who are already fluent in an Asian language with a base-ten numeration system should be encouraged to maintain that fluency, especially in counting. Learning to count in two language systems, especially if the first language is structured on base-ten, should support conceptual understanding in the English number-naming system, which will ultimately become the child's language of mathematics problem solving.

CONCLUSION

We believe that the language of mathematics, particularly the counting numbers, plays a significant role in Chinese, Japanese, and Korean children's superior performance in mathematics. Although this is not the only factor in explaining achievement differences, Asian-language speakers do have a distinct advantage in counting and in understanding what the numbers represent. Understanding these place-value concepts is key to comprehending addition and subtraction with regrouping and also basic to the arithmetic operations of multiplication and division. Asian languages use the base-ten system in their generation of number names, and this system is so transparent that by simply learning the number names from 1 to 10, one can then generate the remaining number names up through 99. Lacking this linguistic support in English, other strategies for learning the basis of the number system should be employed. The activities we have presented attempt to compensate for these language differences by highlighting, through multisensory means, the base-ten structure of the English language numeration system.

REFERENCES

Addison-Wesley. *Addison-Wesley Mathematics: Grade 2.* Menlo Park, Calif.: Addison-Wesley Publishing Company, 1993.

Baroody, Arthur J. *Video Tape Workshops for Teachers: A Cognitive Perspective on Early Number Development.* Paper presented at the annual meeting of the American Educational Research Association, Washington, D.C., 1987.

Bell, Garry, ed. *Asian Perspectives on Mathematics Education.* Lismore, Australia: The Northern Rivers Mathematical Association, 1993.

California Education Policy Seminar. *Lessons in Perspective: How Culture Shapes Math Instruction in Japan, Germany and the United States.* Sacramento, Calif.: The CSU Institute for Education Reform, 1997.

Case, Robbie. *The Mind's Staircase: Exploring the Conceptual Underpinnings of Children's Thought and Knowledge.* Hillsdale, N.J.: Lawrence Erlbaum Associates, 1992.

Cobb, Paul. *Cultural Tools and Mathematical Learning: A Case Study.* Paper presented at the annual meetings of the American Educational Research Association, Atlanta, Ga., 1993.

Freitag, J. *Strengthening Math Learning in First and Second Grade Classrooms.* Bellevue, Wash.: Bureau of Education and Research, 1993.

Fuson, Karen C., Judith L. Fraivillig, and Birch H. Burghardt. "Relationships Children Construct among English Number Words, Multiunit Base-Ten Blocks, and Written Multidigit Addition." In *The Nature and Origins of Mathematical Skills*, edited by Jamie I. D. Campbell, pp. 39–112. New York: North-Holland, 1992.

Gelman, Rochel, and C. Randy Gallistel. *The Child's Understanding of Number.* Cambridge, Mass.: Harvard University Press, 1978.

Kaput, James J. "Notation and Representations as Mediators of Constructive Processes." In *Constructivism in Mathematics Education*, edited by Ernst von Glasersfeld, pp. 53–74. Dordrecht, Netherlands: Kluwer, 1991.

Luriia, Aleksandr Romanovich. *Cognitive Development: Its Cultural and Social Foundations.* Cambridge, Mass.: Harvard University Press, 1976.

Miller, Kevin F., and James W. Stigler. "Counting in Chinese: Cultural Variation in a Cognitive Skill." *Cognitive Development* 2 (1987): 279–305.

Miller, Kevin F., Catherine M. Smith, Jianjun Zhu, and Houcan Zhang. "Preschool Origins of Cross-National Differences in Mathematical Competence: The Role of Number-Naming Systems." *Psychological Science* 6 (1995): 56–60.

Miura, Irene T., Yukari Okamoto, Chungsoon C. Kim, Marcia Steere, and Michel Fayol. "First Graders' Cognitive Representation of Number and Understanding of Place Value: Cross-National Comparisons—France, Japan, Korea, Sweden, and the United States." *Journal of Educational Psychology* 85: (1993) 24–30.

Miura, Irene T., Yukari Okamoto, Chungsoon C. Kim, Marcia Steere, Michel Fayol, and Chih-Mei Chang. "Comparisons of Children's Cognitive Representation of Number: China (PRC), France, Japan, Korea, Sweden, and the United States." *International Journal of Behavioral Development* 17 (1994): 401–411.

Nunes, Terezinha. "Cognitive Invariants and Cultural Variation in Mathematical Concepts." *International Journal of Behavioral Development* 15 (1992): 433–53.

Okamoto, Yukari, Robbie Case, Charles Bleiker, and Barbara Henderson. "Cross Cultural Investigations: The Role of Central Conceptual Structures in the Development of Children's Thought." In *Monographs of the Society for Research in Child Development*, Serial 246, Volume 61, edited by Robbie Case and Yukari Okamato, pp. 131–55. Ann Arbor, Mich.: Society for Research and Child Development, 1996.

Rogoff, Barbara. *Apprenticeship in Thinking: Cognitive Development in Social Context.* New York: Oxford University Press, 1990.

Saxe, Geoffrey B. "Candy Selling and Math Learning." *Educational Researcher* 17 (1988): 14–21.

———. "Developing Forms of Arithmetic Operations among the Oksapmin of Papua New Guinea." *Developmental Psychology* 18 (1982): 583–94

Steffe, Leslie P., Paul Cobb, and Ernst von Glasersfeld. *Construction of Arithmetical Meanings and Strategies.* New York: Springer-Verlag, 1988.

Third International Mathematics and Science Study. *Mathematics Achievement in the Middle School Years.* Chestnut Hill, Mass.: TIMSS International Study Center, 1996.

Vygotsky, Lev Semenovich. *Mind in Society: The Development of Higher Psychological Processes*, edited by Michael Cole, Sylvia Scribner, Vera John-Steiner, and Ellen Souberman. Cambridge, Mass.: Harvard University Press, 1978.

———. *Thought and Language.* Cambridge, Mass.: MIT Press, 1962.

Connecting Mathematics, Language, and Culture for ESL Students

A Guam Classroom

6

Jacquelyn Milman
Ione Wolf
Wilson Tam

Guam is an unincorporated island territory of the United States with a population of approximately 145 000. Located in the Western Pacific, it is a diverse island with an indigenous Chamorro population as well as people coming primarily from Micronesia, Asia, and the United States. The education system is modeled after that of the United States, and instruction is delivered in English. Many students, however, begin school with little or no English. Over one-third of the school population are classified as second-language speakers of English. Many of this diverse cultural and linguistic population also score low in mathematics. Thus, teachers struggle with not only content but also language and culture. The following article examines the cultural and linguistic influences on learning mathematics and explores various strategies for teaching mathematics to children of widely diverse linguistic and cultural backgrounds. In-class practices that have been successful in one sheltered mathematics class for English as a second language (ESL) students at the middle school level are reported.

With the growth of non-English linguistic populations in the United States over the past few decades, language skills have been regarded by many educators as a limiting factor to equal access to classroom instruction. Cummins' threshold theory (in Mestre [1988]) explains this as the necessity for attaining a certain level (threshold) of language proficiency in order for positive benefits to be seen in the academic areas. Non-English speakers quickly attain what he refers to as *basic interpersonal communication skills* (BICS). However, *cognitive academic language proficiency* (CALP) can take five to seven years to develop. This may explain why many non-native speakers appear to have good ability in English communication yet fail miserably when placed in the mainstream classroom.

While the relationship between language skills and academic performance is not yet clearly understood, mathematics achievement appears to be one area that depends on a strong relationship with language skills. Studies point to the strong effects of language on mathematics achievement, and a positive relationship between mathematics achievement and verbal ability has been documented in the general population (Cocking and Chipman 1988). This may be a key to the high incidence of low mathematics achievement among language-minority students.

By ethnicity, Guam's school population is approximately 53 percent indigenous Chamorro, 27 percent Asian, 7.5 percent Micronesian, 8 percent Caucasian, 2 percent African American, and 2.5 percent other. One-third of this population has English as a second language (ESL). Although a breakdown by ethnicity was not available, overall mathematics scores have consistently remained below the 40th percentile on norm-referenced tests, with the exception of the third grade (Guam Department of Education). To date, no research has been done among this population to ascertain if low English proficiency can be said to be a cause of low mathematics scores, but the fact that many limited-English-proficient students have difficulties in mathematics is a cause of concern, and ways are being sought to teach them more effectively.

A REVIEW OF THE LITERATURE

A search of the literature reveals three primary areas of concern—(1) the influence of culture on mathematics achievement, (2) the relationship of language and mathematics, and (3) the language of mathematics—and their implications for teaching.

The Influence of Culture on Mathematics Achievement

There are many studies among the general population that offer support for the intrinsic relationship between language background and cognitive development. Saxe (1988, 1991) extends the case to language background and *mathematical* cognition. He suggests one should study cultural supports for mathematical development and look at how children use different cultural backgrounds to cope with the school mathematics curriculum. Such study, he contends, will offer insights regarding the sources of the language-minority child's success or failure in mathematics. This stance is supported by several researchers (Bickmore-Brand 1990; Gawned 1990; Charbonneau and John-Steiner 1988; Gay and Cole 1967).

Charbonneau and John-Steiner (1988) look at several studies that examine cultural patterns of experience in learning mathematical concepts and their relationship to mathematical achievement in school. They strongly recommend that we make connections between informal and formal (school) learning and use cultural activities that support mathematical concepts—hands-on, experiential learning that is structured to reflect the culture and interactional patterns of the community from which the child comes. Activities pertinent to Pacific cultures include allowing children to observe and imitate adults, making use of oral tradition by having students verbalize, using concrete examples familiar to students (story problems related to fishing, for example), employing small, cooperative groups, and encouraging peer consultation.

The Relationship of Language and Mathematics

Limited-English-proficient students are often mainstreamed in math on the erroneous assumption that math is language independent since it involves numbers (Spanos et al. 1988; Chamot and O'Malley 1994). Spanos et al. (1988) propose a language approach to the teaching of mathematics. Mestre (1988) likewise suggests teaching mathematics and language together. Verbalizing problem-solving procedures (Chamot and O'Malley 1994) and cooperative activities (Chamot and O'Malley 1994; Brunovsky 1991) have also been suggested. Cooperative learning activities, says Brunovsky (1991), provide a bridge from one language to the other by providing a social support mechanism for success. They actively engage students, provide opportunity for creative thinking, and allow for peer assistance.

Bickmore-Brand (1990) looked at the implications of language arts research for mathematical teaching and identified seven areas that could also be relevant to mathematics: context, starting point, modeling, scaffolding, metacognition, responsibility, and community. Context must be meaningful and complete, not merely isolated textbook problems. We must start where the learner is and not view him/her as deficient. The teacher must be a model and demonstrate the use of mathematics in life. *Scaffolding* refers to building upon what is already known and challenging the learner to increase his/her capacity. The process must be made explicit and students must develop responsibility for their own learning. The classroom environment should be supportive and risk-free.

Chamot and O'Malley (1994) identify language skills that are required in mathematics. Listening skills are needed to understand oral numbers, oral word problems, and explanations that have no concrete referents. Reading is required to read and understand specialized vocabulary, textbook explanations, and word problems. Speaking is necessary to answer questions, ask questions for clarification, explain problem-solving procedures, and describe math applications in other areas. Writing is needed to write verbal input numerically, show solutions for word problems, and combine words for number sentences.

Bye (in Dale and Cuevas [1987]) analyzed the discourse in mathematics texts and found them to be highly conceptual. Mathematics requires up-and-down and left-to-right eye movements, adjustment of reading rate to accommodate the specialized language, and multiple readings. It uses many symbolic devices, such as charts and graphs, and contains much technical language. These discourse features tie in to reading ability (Dale and Cuevas 1987). A second-language student who is a weak reader will also have problems in mathematics. He or she must have both language proficiency and mathematical skills and be able to integrate the two. Metacognition, curiosity, and a willingness to investigate the unknown are necessary for mathematical thinking.

The Language of Mathematics

Mathematics vocabulary is highly specific and lacks the redundancy and paraphrasing that often aid comprehension in other types of language (Chamot and O'Malley 1994; Dale and Cuevas 1987). Structures that are commonly used in the language of mathematics seldom appear in other content areas or even in English as a second language (ESL) classes, so students have little opportunity to gain experience or practice with them. Thus, the language is best learned through context with an emphasis on communicating the concepts, processes, and applications of mathematics. Cuevas (cited in Dale and Cuevas [1987]) developed a Second Language Approach to Mathematics Skills (SLAMS) that incorporates language-development activities into the mathematics class. Conversely, problem-solving activities should also be developed in the mathematics class that promote second-language acquisition.

Short and Spanos (1989) and Secada and Carey (1989, 1990) echo the need for language-sensitive content instruction and a focus on the language of mathematics. Secada and Carey (1989, 1990) give several suggestions for achieving these objectives. One is to use the children's own experiences or to create a class experience (Dixon and Nessel 1983). The latter is a technique, termed the Language Experience Approach, often used to teach reading. The following two types of instruction are recommended: Active Mathematics Teaching (AMT), a form of direct instruction that conveys large amounts of highly structured material to beginners, and Cognitively Guided Instruction (CGI), which focuses on student thought processes as they solve problems.

Mathematics involves problem solving, and problem solving uses the language of reasoning, in which complex sentences (those with two main verbs) and coordinating or subordinating conjunctions are most frequently used to express the meaning of relationships (Gawned 1990). The language of mathematics is symbolic, content specific, and descriptive (describes observations). It uses labels, attribute nouns, and noun-phrase constructions. It is also procedural (to explain how and why) and uses language sequentially to enumerate steps and explain and justify outcomes.

Mestre (1988) discusses forms of language proficiency and emphasizes that learners need both general and technical language in mathematics. Dale and Cuevas (1987) point to the frequent use of logical connectors (*then*, *if*, etc.) in mathematics. The student must be able to make inferences in order to identify a referent's variable(s).

MacGregor (1990) says that mathematics problems have the structural characteristics of a riddle. They involve finding the fundamental reference and how other elements relate to it. They require syntactic awareness, analytic reading skills, and the use of prepositions. Mathematics also involves writing that puts thinking to work by expressing ideas on paper and helping to discover and to clarify. And it reveals misconceptions and gaps in understanding. For second-language speakers, vocabulary and word order are easier to learn than auxiliary words that designate specific relationships between parts of sentences.

WHAT WORKS IN ONE GUAM CLASSROOM

The ninth grade middle school classroom used for this article is typical of sheltered classes on Guam for second-language students. The students were ten males and six females at the intermediate level of English proficiency, who were of Chamorro, Chuukese, Yapese, Palauan, Filipino, and Vietnamese ethnicity. A daily journal was kept by the instructor to record problems, strategies and techniques, and their results.

Sheltered classes contain only ESL students. They are taught by ESL teachers (termed *language other than English* [LOTE] by the Guam Department of Education) in the subject areas of language arts, mathematics, science, and social studies. In a sheltered mathematics class, the teacher integrates language and mathematics learning in order to enhance students' language skills while studying mathematics. A typical week in the classroom of this article consists of the following:

Monday   *Presentation and lecture.* Students spend time reading the lesson content, and the teacher explains the lesson.

Tuesday   *Practice of skills related to the lesson.* Students have the opportunity to practice their math skills by doing math problems.

Wednesday   *Presentation of learning strategies necessary to solve word problems.* Students are taught four steps to follow in attacking word problems: (1) identify the question(s), (2) collect and analyze the data, (3) choose the operation (add, subtract, multiply, or divide), and (4) solve the problem.

Thursday   *Integration of language learning and mathematics through writing.* The students are taught to write simple word problems relating to the lesson. This exercise enhances the development of higher order thinking as well as writing skills.

Friday   *Evaluation of students performance and practice of the multiplication table.* The teacher uses a quiz to help students retain what they have learned and improve their multiplication skills.

By keeping a daily journal, the teacher was able to analyze difficulties encountered by students and develop lessons to respond to the difficulties. An example of one journal entry follows:

A review of adding and subtracting whole and decimal numbers was conducted in class today. It is very common among the students to make mistakes while subtracting. Students often subtract the smaller number from the larger number. A typical example of this error is

$$
\begin{array}{r}
70 \\
-28 \\
\hline
58
\end{array}
$$

Since 8 is greater than 0, students subtract the smaller number, 0, from the larger number, 8.

Another common mistake occurs in aligning the decimal points when adding decimal numbers. Instead of aligning the decimal points, they often align the numbers.

$$
\begin{array}{r}
1.5 \\
+\ 2.24 \\
\hline
2.39
\end{array}
$$

It became readily apparent that students were performing the mathematical procedures without understanding them. What was not so obvious was why comprehension was lacking. Culturally, the students were from cooperative societies. The Micronesians come from an oral tradition. Children learn from observing and imitating adults. Many tend to be very shy and reluctant to speak up in class. Keeping these characteristics in mind, the teacher set about to resolve some of their difficulties in mathematics.

Mathematical Background

One of the first problems tackled was that of multiplication. Almost all students were having difficulty with the times table beyond five. Several strategies were tried. One which proved particularly helpful was to draw pictures of problems, presenting multiplication as a system of rows and columns. For example, for the problem 3×5, the teacher drew 3 rows of circles with 5 columns of circles in each row.

$$
\begin{array}{ll}
& \text{o o o o o} \\
\text{rows} & \text{o o o o o} \\
& \text{o o o o o}
\end{array}
$$

columns

After the teacher modeled the procedure, students were asked to "draw" problems in front of the class, in groups, in pairs, and individually. Heretofore meaningless numbers were represented by fish, coconuts, or breadfruit trees. Context was added by creating story problems to accompany the numbers. Showing the students the problems through drawings opened their minds to a higher level of reasoning. It was the first time some of the students understood what multiplication was beyond a rote memorization task.

A further problem with multiplication was that students were only able to recite the times table in a fixed pattern; they lacked practice in using it randomly. Flash cards were used, and the class divided into teams to make a competition of it. Only two to four numbers were tackled at a time until the students became familiar with the times table.

Another technique employed for multiplication was to have the students construct a multiplication table using the square floor tiles in the classroom. A table for multiplication from two through nine was constructed. The boundary for the table was outlined with black electrical tape. Numbers were also cut from the tape and attached to the floor. This was then used both as a game and as a life-sized tool for problem solving such things as perimeter and area.

To play the game, students were divided into two teams. The teacher would call out a multiplication question. One student from each team would try to locate the correct answer by standing on the tile. The first to locate the correct answer won a point for his or her team. Student interest was thus heightened as they improved their speed and accuracy in solving multiplication problems.

To insure that this skill was not forgotten, one day per week was at least partially spent reviewing the multiplication tables in some way, whether it be practice with problems, flash cards, or a game. This helped keep skills sharpened and ready for use in more advanced material.

Language

To improve reading skills, one day per week was set aside to work specifically on language skills. Acknowledging the cultural preference of these students for verbal activities, the teacher conducted read-alouds of word problems, with the students reading after him or one-by-one, followed by explanation of any unknown vocabulary. Mathematical terms such as *product, quotient, sum,* and *difference* were often problematic in the beginning.

The reading was followed by questions about the problems to sharpen the students' reasoning skills. Drawings were frequently used to enhance understanding by illustrating what the words were saying. Concrete objects and other visuals were also used.

Students were next asked to tell, in writing, what the question asked them to do. A poor background in mathematics and a difficulty in writing and expressing thoughts made this task extremely hard at first. Therefore, the instructor gave the students a four-step process to guide them in tackling word problems:

1. Write what the question is asking. (The instructor felt that this was something they could accomplish that would thus boost self-esteem, something which was often found lacking among the second-language students.)
2. List all information that might help solve the problem.
3. Identify which operation is needed to solve the problem.
4. Solve the problem.

This method was also applied to number problems by having students write the words for the question, give information, and state what operation should be used. Later, rather than state the question, they were asked to create questions of their own by using the numbers in the mathematics problems. Using the drawing method described above aided students in this task.

Initially, there was less concern about how well students could express their understandings on paper. Focus was on being able to analyze the problems and arrive at an answer mentally. Once they could do this, then transfer of their understanding to paper became the focus. Even those students who continued to have difficulties exhibited more confidence.

Another technique was to put the students into small groups and allow them to discuss the problems and assist each other. Students who were often reluctant to perform alone in front of the class would willingly participate in a small group. Cooperative learning tasks were used to encourage participation and exercise metacognitive skills.

Toward the latter part of the term, students were requested to write their own mathematics problems to both enhance their English writing skills and to practice their mathematical reasoning skills. Three to four numbers would be chosen randomly, then labeled (such as 3 birds, 5 fish, etc.). Then questions would be formed using the numbers chosen. Students were required to solve the problems they created to test the validity of their questions. Writing was the most difficult aspect of this procedure for them. Although they were working with fractions and decimals in their lessons, questions were, at first, limited to using whole numbers until they became more familiar with the practice. Then questions were tied to lessons and, if working on place value, for example, they were required to make questions relating to place value.

IN CONCLUSION

Many of the techniques found to be successful with the particular group of students studied are techniques suggested in the literature for culturally and linguistically diverse classrooms in general. Using concrete objects, visuals, manipulatives, drawings, and graphics helped the students to visualize and make sense of the numbers and symbols, concepts and ideas, and rules and formulas they were being asked to learn. Using flash cards and games helped improve speed and accuracy in solving multiplication problems. Teaching them the four-step process for attacking problems allowed them to experience success in solving word problems. Writing word problems improved not only mathematical skills but also English-writing ability. Real examples from their own background and experience gave significance and personal meaning to the role of mathematics in their own world. Peer consultation and cooperative activities increased confidence and self esteem. Allowing verbalization before reading and writing about problems aided comprehension and developed and maintained metacognition and language skills.

More experimentation needs to be done regarding individual ethnic groups within the larger classroom population. It is believed that much about the various cultures and the means by which mathematics are incorporated could be explored and used for more effective teaching. There is much untapped knowledge to be gleaned for the teacher who is willing to investigate and make use of the knowledge in the classroom.

REFERENCES

Bickmore-Brand, Jennie. "Implications from Recent Research in Language Arts for Mathematical Teaching." In *Language in Mathematics*. Portsmouth, N.H.: Heinemann, 1990.

Brunovsky, Kathy. "Cooperative Learning in Mathematics with ESL Students." *Elementary ESOL Education News* 13, no. 2 (Winter 1991).

Chamot, Anna. U., and J. M. O'Malley. *The CALLA Handbook: Implementing the Cognitive Academic Language Learning Approach*. Reading, Mass.: Addison-Wesley Publishing Company, 1994.

Charbonneau, Manon P., and Vera John-Steiner. "Patterns of Experience and the Language of Mathematics." In *Linguistic and Cultural Influences on Learning Mathematics*, edited by Rodney R. Cocking and Jose P. Mestre, pp. 91–100. Hillsdale, N.J.: Lawrence Erlbaum Associates, 1988.

Cocking, Rodney R., and Susan Chipman. "Conceptual Issues Related to Mathematics Achievement of Language Minority Children." In *Linguistic and Cultural Influences on Learning Mathematics*, edited by Rodney R. Cocking and Jose P. Mestre, pp.17–46. Hillsdale, N.J.: Lawrence Erlbaum Associates, 1988.

Dale, Theresa Corasaniti, and Gilberto J. Cuevas. "Integrating Language and Mathematics Learning." In *ESL through Content-Area Instruction: Mathematics, Science, Social Studies*, edited by JoAnn Crandall, pp. 9–54. Englewood Cliffs, N.J.: Prentice Hall Regents, 1987.

Dixon, Carol N., and Denise D. Nessel. *Language Experience Approach to Reading (and Writing)*: LEA for ESL. Hayward, Calif.: Alemany Press, 1983.

Gawned, Sue. "The Emerging Model of the Language of Mathematics." In *Language in Mathematics*, edited by Jennie Bickmore-Brand, pp. 1–9. Portsmouth, N.H.: Heinemann, 1990.

Gay, John, and Michael Cole. *The New Mathematics and an Old Culture: A Study of Learning among the Kpelle of Liberia.* New York: Holt, Rinehart and Winston, 1967.

Guam Department of Education. Unpublished student data.

MacGregor, Mollie. "Reading and Writing in Mathematics." In *Language in Mathematics*, edited by Jennie Bickmore-Brand, pp.100–108. Portsmouth, N.H.: Heinemann.

Mestre, Jose P. "The Role of Language Comprehension in Mathematics and Problem Solving." In *Linguistic and Cultural Influences on Learning Mathematics*, edited by Rodney R. Cocking and Jose P. Mestre, pp. 201–20. Hillsdale, N.J.: Lawrence Erlbaum Associates, 1988.

Saxe, Geoffrey B. *Culture and Cognitive Development: Studies in Mathematical Understanding.* Hillsdale, N.J.: Lawrence Erlbaum Associates, 1991.

———. "Linking Language with Mathematics Achievement: Problems and Prospects." In *Linguistic and Cultural Influences on Learning Mathematics*, edited by Rodney R. Cocking and Jose P. Mestre, pp. 47–62. Hillsdale, N.J.: Lawrence Erlbaum Associates, 1988.

Secada, Walter G., and Deborah A. Carey. *Innovative Strategies for Teaching Mathematics to Limited English Proficient Students.* Program Information Guide Series, Number 10. National Clearinghouse for Bilingual Education (Summer 1989).

———. *Teaching Mathematics with Understanding to Limited English Proficient Students.* Urban Diversity Series No. 101. ERIC Clearinghouse on Urban Education Institute on Urban and Minority Education (October 1990).

Short, Deborah J., and George Spanos. "Teaching Mathematics to Limited English Proficient Students." *ERIC Digest* (November 1989). Washington, DC: Center for Applied Linguistics.

Spanos, George, Nancy C. Rhodes, Theresa Corasaniti Dale, and JoAnn Crandall. "Linguistic Features of Mathematical Problem Solving: Insights and Applications." In *Linguistic and Cultural Influences on Learning Mathematics*, edited by Rodney R. Cocking and Jose P. Mestre, pp. 221–40. Hillsdale, N.J.: Lawrence Erlbaum Associates, 1988.

Positive Strategies That Focus on Cultural Roots Help Asian Students Adjust to American Schools

7

Mangho Ahuja

Kayo Matsushita

Tamela D. Randolph

Note to the Reader: The notations [A] and [M] refer to the first two authors, respectively.

There is a general impression in the academic world that Asian students are smart and that they do well on tests (Suro 1994). They are also well behaved, respectful to teachers, and display fewer disciplinary problems than students of other ethnic backgrounds. What is the underlying source of their motivation for such performance and behavior? How much do the family and cultural values contribute to their success in school? Does the juxtaposition of the two cultures, Asian and American, help or hinder the learning process?

This paper contains a narration of some conversations, incidents, and stories that sketch the contrast between the educational and cultural systems of some Asian countries and the United States. It tries to explore the hearts and minds of Asian students who are faced with the conflicting demands of the social and cultural traditions of the two cultures. It explains how the differences in the two cultures cause stress on the parents of Asian students and their American teachers. It emphasizes the need to understand different cultures. Lastly, it offers some suggestions for developing and employing positive teaching strategies for a multicultural classroom. Our aim is not to justify, glorify, or criticize any culture or educational system but to explain the cultural roots of the behavior of Asian students. As Carol Archer puts it, "Why do they do what they do? And, how do they do what we do? In other words, why are we different and, finally, how are we similar?" (Archer 1994, p. 82).

ELEMENTARY SCHOOL MATHEMATICS IN INDIA

When I [A] was attending elementary school in India, sometimes my Uncle Hari would visit my home. After making preliminary inquiries about everyone's health, he would approach me and say, "So how is school these days? You say you are in the second grade. Let us see what you have learned so far." Then he would rapidly fire off some questions: "What is 7 times 12? What is 8 times 17?" and so on. In those days (1940s), students were expected to learn the multiplication tables from 2 through 20 by the third grade, so these questions were not unreasonable. If I answered the questions correctly, he felt satisfied with my progress in school.

Soon after I came to the United States, I [A] went to a local grocery store. I bought 12 items of the same kind at 7 cents apiece. At that time (1960s), the girl at the check-out counter had to punch 7 cents 12 times. While she was punching and counting, I said, "84 cents." "Wow! You are a real math wizard! How did you know that?" she said. As I was leaving the store, I was reminded of Uncle Hari in India. For Uncle Hari, my mastery of the multiplication tables meant that I was making satisfactory progress. To the girl at the checkout counter, it meant something else —I was some kind of a math genius or perhaps a weird or an abnormal person.

It is interesting to note that while elementary education in India is known to be more demanding than that in the United States, most Indians are very critical of the Indian system and prefer American education. They say:

> The education in India is so bookish! You pass exams in the First Division (something similar to getting an A), but you don't learn a thing. It is mainly rote memorization; you don't learn anything useful or practical. We learn only to pass exams.

This criticism is valid because passing exams at various levels forms the structural framework of most Asian educational systems. However, in addition to preparing students for exams, Asian schools must provide the cultural roots and build the moral fabric of the growing mind. Teachers must teach and enforce the rules of social behavior. That is why the society puts a high premium on prestigious schools in Asian countries.

HIGH REGARD AND RIGID RULES

Because learning is usually measured by a students' performance on important exams at various stages, the preparation for a good education begins as early as kindergarten or nursery school. To ensure the high quality of entering students, nursery schools have their own elaborate entrance exams. Even the parents of the prospective student must appear for an interview with the principal of the school. The Indian parent would proclaim, "My child goes to Blue Bonnet School," with almost the same pride as an American parent might say, "My child goes to Yale." The prestigious nursery schools have good facilities, but their cost is enormous. This favors the wealthy and eliminates the average Indian family. In Asian countries, the education of a child involves significant cost and sacrifice from the parents. After all, the entire career of a young Asian student in India, Japan, and China hinges on his or her education.

Most Asian schools have rules governing the behavior, dress, and activities of its students. In many Japanese schools, students entering middle school are given a "blue book" that describes in great detail the rules regarding such items as the length of their hair, the length of their skirts, and the color of their socks. The book comes complete with diagrams explaining the rules precisely. Boys' hair should not touch the back of the ears. For girls, the nature, length, and color of each part of their attire is clearly prescribed. Neither girls nor boys can wear earrings. For any infraction of the rules, the child is scolded and punished by the teacher. Every aspect of the students' appearance must fall within the prescribed norms as set forth in the blue book. The Japanese emphasize uniformity and homogeneity. Most schools have their own school uniform. It is no surprise that to an American traveler, all Japanese students look alike. One can understand the bewilderment of Japanese teachers and students when they visit American schools. While the Japanese culture cultivates homogeneity and uniformity, we encourage individualism. While they encourage obedience to authority, we admire the ability to question, argue, and criticize. The principle of sameness there is replaced here by the dictum "dare to be different."

COLLEGE ENTRANCE EXAMS

In most parts of Asia, the performance on competitive exams determines the entire career of a young student. So, generally speaking, the educational system is geared to passing and getting high scores on these exams, and the focus on learning is somewhat blurred. In India, the exams come at the end of high school, and only those scoring high enough marks are eligible for admission into reputed colleges and professional schools. For all Indians, the two most desirable careers are engineering and medicine.

In China and Japan, the most important exams are the College Entrance Exams. The focal point of the Japanese education is the entrance exams (Cummings 1979). From the first day a Japanese child begins school, the child and the parents start worrying about the entrance exams at different levels. The prestigious jobs go to the graduates of elite universities—Tokyo, Kyoto, Waseda, Keio, etc. The reputation of a good high school is measured by the percentage of its graduates passing the College Entrance Exams of the top universities.

To enter a prestigious educational institution at any level, one must pass its entrance exam at that level. Thus, there are entrance exams for kindergarten, junior high, senior high, and finally the College Entrance Exam for entering the university. For a Japanese student, the College Entrance Exam is the ultimate challenge of a lifetime. As Carl Becker (1990) put it, "[c]ollege entrance becomes the most important testing time in the life of the person, determining his or her future employment, social and possibly marital status, life income, etc. It is only natural that parents should want their children to receive the best possible jobs and salaries—for which the name of the university their children enter becomes critical. It is this system which creates the 'pressure cooker environment,' the so-called 'examination hell' from which so many students suffer breakdowns and psychoses each year" (p. 434).

On failing to gain admission into the university of their choice, the Japanese students must then try somewhere else. The Japanese universities may be classified as national (public), private, metropolitan (located in a large city like Tokyo), or local. The national universities (formerly the seven Imperial universities) have limited seats and can only accommodate the best students. Now, however, some private metropolitan universities, especially Waseda and Keio also have good reputations. Private universities are costly. Thus, the parents may decide to send their child to America for a variety of reasons—to receive an American education, to learn the English language, to avoid the entrance exams, or perhaps because socially it is much more acceptable to get an education at an American university than at a non-prestigious Japanese university. Monetarily, the cost of education in the United States is usually about the same as at a private university in Japan. These are some of the reasons why we find Japanese students in American classrooms.

COMING TO AMERICA

On arrival in the United States, Asian students find the learning environment here quite unfamiliar. Here, besides passing tests, one is involved with writing papers, making book reports, participating in class discussions, making class presentations, sending assignments by e-mail, putting problems on the board, doing projects, and working together in teams. While the Asian school systems focus on exams, the American system, some experts believe, unduly emphasizes extracurricular activities. As a result, American teachers quite often hear the following remarks. "My dad does not want me to join the choir. He says it is for young ladies." "My mom does not want me to wear tank tops or shorts for tennis. We are Moslems." "My mom does not like me to go on overnight trips for speech or debate. She gets awfully worried about my safety." "My dad does not want me to join in the school play. He says I should join a math or a science club instead." It appears that the diverging forces of the two cultures meet face to face at the doorstep of the American teacher's classroom. What is an American teacher supposed to do? Some courageous teachers have shown a remarkable ability in arguing the student's case with the parents.

First they make an appointment to meet the parents. Feeling almost like they are playing the role of Dr. Henry Kissinger, they enter the house of Mr. Shah, Dr. Afzal, Mr. Nagata, or Mr. Chow. Over a cup of Asian tea and between the complimentary remarks about the Asian artifacts in the house, they are able to put the parents at ease. "Oh, Nisha will be safe on overnight trips. I will personally look after her." Asian parents can hardly ignore the request of their son or daughter's teachers. They usually give in. Such teachers deserve our respect and praise because they are the true architects who are building the bridges connecting the diverse cultures.

RETURNING TO JAPAN

What happens when Japanese students return home? The culture conflict continues. The returnees, called "Kikokushijo" in Japanese, feel changed or "transformed" in some way. Many of these changes are not acceptable in Japan. "Some transformations are short-lived physical changes in hair and clothing; some are visible behavioral changes in how they walk or move their arms or faces. The more serious changes are in interpersonal styles and expressions, changes that make them seem 'un-Japanese.'" (Kidder 1992, p. 383). One returning young Japanese woman was very specific and said, "Many friends who went to the U.S. came back and walked with their toes out, like a man, and it looks very cool, not girlish ... and with big steps. Girlish is with toes in or parallel, and not much space between the steps" (Kidder 1992, p. 386).

The returnees obviously failed to learn the fine nuances of their language. In America, we use the same words irrespective of the age of the person to whom we are speaking. The Japanese use special language terms when they talk with their elders. They call it "keigo," which means showing respect to one's elders. There are differences in behavior, speech, and dress. Japanese say that in the U.S., people are "too direct"; in Japan they try to speak "rounder" (p. 387). The returning Japanese students now question every aspect of their educational as well as social system. They have learned to argue and ask questions directly and not accept authority blindly. Such behavior is quite unlike that of traditional Japanese students who are discouraged from questioning established rules. The American teen culture has a powerful influence on the young mind; this greatly worries the parents. Obviously, the Japanese students who have lived for some years in the U.S. hate to go back and lose all of the freedom to which they have been accustomed.

CULTURAL DIFFERENCES

The Japanese students enter American universities to learn western culture, to study English, and to receive an American education. How do they perform at the university level? In the Asian culture, "good" students behave politely in school and do not complain or argue. In mathematics classes, their focus is more on learning how to solve problems. So, when they come to the U.S., their class participation is minimal. Japanese students, when left to themselves, tend to huddle together in the back of the room, do not share their views with others, rarely ask questions, and remain passively in the background. Wakako (W) (girl) and Susumu (S) (boy) were two such students from Japan in my [A's] college algebra class. They used to enter the class quietly, find the remotest corner of the classroom to sit, and try to make themselves invisible. After some thinking, I found a way for them to interact with the rest of the class. I had prepared assignments for each student for an oral presentation in the class. Other students were to talk about Gauss or Newton or Omar Khayyam, but the assignment for W and S was to talk about the Japanese high school system. On the day of the presentation, they appeared very nervous. In lilting voice and with some emotion, W talked about her junior high and S about his senior high school years. The entire class fell silent and listened intently to their various episodes that described the academic training and the strict discipline of their high school system. At the end of their presentations, the students compared the Japanese system with the American system. From then on, the students would often refer to this or that episode of S and W and joke with them. S and W too became more alive and responded jovially to their comments. Now it did not matter where W and S were seated, for they were a part of the class.

While teaching a college algebra class, which consisted only of newly arrived Japanese students, I [A] noticed something unusual. Whenever I asked one of them a question, no matter how simple, he or she would first mumble something to others and answer me only after consultation. Naturally, I thought this was cheating. Later, my co-author [M], as well as my colleague who had spent many years in Japan, explained this behavior to me. The Japanese students are not very confident of their English. Also, they are very "correct-answer oriented" and would be ashamed if they made a mistake. Finally, as is well known about the Japanese system, the entire group or team shares the responsibility for the final outcome. This is why the Japanese students who came as a group liked to consult each other. They were trying to translate their answer from Japanese to English, and to make sure that their final answer in English was correct, consultations made them feel more comfortable.

While visiting a junior high school mathematics class in the United States, I [M] was shocked by the following observation. When the teacher asked who did not get the right answer, a few hands went up. One of them spoke, "Ms. S, I did all that but got a different answer. What did I do wrong?" I was shocked. Japanese students would never disclose their mistakes in public. They would be ashamed to say they got a wrong answer.

THE NEED FOR MULTICULTURAL KNOWLEDGE

The following story (Ferguson 1987, p. 29) is very illuminating and makes a convincing argument for the need to understand different cultures. A Chinese student did something wrong. His American teacher asked him to approach the desk. While the teacher admonished him, the student kept his head down with his eyes glued to the ground. The teacher got very angry and said, "Look at me when I am talking to you!" But the student kept his head down. The teacher became more angry and sent him to the principal for more punishment. According to Chinese culture, one is not supposed to look into the eyes of an elder while being spoken to or being rebuked. It is very disrespectful. According to his culture, the student's behavior was very normal and respectful. The knowledge of different cultures helps us understand the behavior of our multicultural students and explains the reason why they do what they do. And, as we have seen above, lack of such knowledge can sometimes lead to grave errors.

LANGUAGE DIFFICULTIES

Language is a cause of problems for some of our Asian students. Ask either a Chinese or a Japanese student, "Don't you want to go to the mall?" If the student answers "Yes," he means, "You are right. I do NOT want to go to the mall." If the student answers "No," he means your supposition is wrong, and he DOES want to go to the mall. This is a good example to discuss in class while explaining "negation" in teaching Boolean algebra because it shows how the logical language sometimes differs from the everyday language. While this example only shows the linguistic problem of the use of the word "no," there is a cultural side too. My [A's] Japanese language teacher explained that the Japanese try to avoid the use of the word "no" and find different ways to do so. If one goes to a store and asks for hamburger buns, and if the store is out of hamburger buns, the storekeeper will never say, "No, we don't have hamburger buns." He will skirt around the issue and ask, "Would you like to have some hot dog buns?" If one persists, "Do you have some hamburger buns?" one must not expect to hear, "No, we don't." More likely one would hear, "Excuse me, are you sure you don't want hot dog buns?"

In mathematics too, different languages and cultures can cause a variety of interesting situations for both the students and their teachers. Here are a few examples from the Japanese language.

1. *Reading a fraction.* While teaching the college algebra class mentioned earlier, I [A] made an interesting observation. Every time the Japanese students mentioned a fractional expression like a/b, they would start saying "b over ...," then change their minds in the middle of the sentence, start all over again, and correctly answer "a over b." When this happened many times, I asked them, "How do you say 3/5 in Japanese?" The answer was "Go bun no san", which means "Of five parts take three". Thus "a over b" would be read as "b bun no a." In other words, while reading a fractional expression, the Japanese start with the denominator first and the numerator later.

2. *Reading a large number.* In writing a large number, we divide the digits into groups of three, e.g., 1 762 456 389. We also read it in groups of three as 1 billion, 762 million, 456 thousand, 389. The Japanese write large numbers just as we do, but when it comes to reading them, they mentally rearrange the digits in groups of four like 17 6245 6389 and read it as 17 oku 6245 man 6389, where oku = 10^8, and man = 10^4. This is because originally the numbers were written in Kanji numerals and were read using the denominations *oku* and *man*, as if they were grouped like 17 6245 6389. With the modernization of Japan in the Meiji era, they switched from Kanji numerals to Hindu-Arabic numerals and also adopted the style of writing a large number in groups of three. But although the writing style changed, the language remained the same. The Japanese language has no words equivalent to a million or a billion. The switching back and forth between the groups of three and groups of four is difficult if it is done mentally. Many Japanese businessmen write down the number and then rearrange the commas to convert from one system to the other.

3. *Reversed word order.* In learning arithmetic, we usually think of "3×5" as 5 things taken 3 times or as 3 bags containing 5 apples each. But in the Japanese language "3×5" is read as "3 no 5 bai" (no = of, bai = times), which means 3 things taken 5 times or an equivalent of 5 bags containing 3 apples each. In other words, we write the multiplier first and the multiplicand second, whereas their order is reversed in the Japanese language. In algebra, we read "$2a$" as "2 times a," keeping the word order of 2 and a the same. But a Japanese would read "$2a$" as "a no 2 bai," or equivalently "take a two

times", thereby reversing the order in which 2 and *a* appear. Because the English language word order is the same as in the mathematical language, it is easy for us to translate the phrase "twice of *a*" into "2*a*." But, for a Japanese child who is learning the basics of algebra, switching from the phrase "*a* no 2 bai" to "2*a*" is quite confusing.

4. *Math language and everyday language.* We read the function notation $f(x)$ as a "function of *x*," but a Japanese student will read it as "*x* no Kansuu" (no = of, Kansuu = function), thereby reversing the word order of *x* and Kansuu, as we have seen in the previous examples. A more important fact is that learning the concept of the mathematical word "function" is easier for us, because "function" is also a word in everyday English language with almost the same meaning. However, the word Kansuu has no meaning in everyday Japanese language. Therefore, to understand $f(x)$, the Japanese students have to learn both a new word and a new concept.

A mathematics educator does not have to be an expert in linguistics, but a watchful teacher would be able to recognize some of the language problems of his multicultural students and possibly use them to his advantage in teaching mathematical concepts. For example, a teacher could ask the class what "3×5" means. Is it "3 bags containing 5 apples each" or "5 bags containing 3 apples each"? He can then gently guide the students to the commutative property.

CULTURAL AND PARENTAL DEMANDS

For students from the Southeast Asian countries, such as India, Pakistan, and Sri Lanka, who are of British heritage, language is not a major problem. The American teachers would notice that these students have other problems caused mostly by the conflicting demands of the two cultures. These demands come mainly from the parents and surrounding community of friends and relatives. The scope of the problems varies with the degree to which the Asian parents have adjusted to the American culture, language, and customs. Without the parents' acceptance of American culture, even a common word like "hi" can be misinterpreted. A young Indian student came home from school and greeted his parents by saying, "Hi, Mom; Hi, Dad." His father admonished him because he disapproved of the casual tone of "Hi, Mom; Hi, Dad." These western greetings do not carry sufficient respect in the eyes of some Asian parents. He should say "Pitaji" (respected father) and "Mataji" (respected mother).

Asians, especially Indians, demand much from their children. When a young premedical student fails to gain admission into a medical program in the United States, he or she is sent back to India where, with the help of a huge donation, the student is admitted to a medical college. The parents' dreams play a major role in deciding the career choice of the youngster. At one time, the "Letters to the Editor" column of the popular Indian weekly *India Abroad* contained a continuing argument between the elders who supported the parents' right to "guide" their children and the young adults who felt chained to the old Indian values while living in the United States.

CONCLUSION

When my [A's] daughter went to kindergarten (in early 1970s), she kept herself busy cutting and pasting paper day after day. Finally, one day, I went to the school and asked her teacher, "When is she going to learn A, B, C, and do arithmetic?" The teacher replied, "Don't be too hasty, Dr. A! The child must learn to adjust to the school's surroundings, learn social skills, learn coordination skills.

The rest of the learning will come slowly." This was not a satisfactory answer to me. My response was, "My daughter knows all that already, and she has no social problems. The sooner she learns to read, write, and do arithmetic, the farther she will be in her schooling." Such views are typical of Asians even today.

The Asian parents have a different view of "learning" and "education" in general, and they strongly believe in the values of their own culture. The Asian children have to obey the traditions of the Asian culture at home but must learn to function and compete in the American environment at school. The American teacher, to be an effective teacher, must recognize the problems of his or her students and understand the cultural roots of their behavior. Sometimes he or she needs to develop innovative teaching strategies to use the environment of his or her multicultural classroom.

REFERENCES

Archer, C. "Managing a Multicultural Classroom." In *Learning across Cultures,*" edited by Gary Althen. pp. 73–88. Washington, D.C.: National Association of Foreign Student Advisors, 1994.

Becker, Carl. "Higher Education in Japan: Facts and Implication." *International Journal of Intercultural Relations* 14 (1990): 425–47.

Cummings, Walter K. "Expansion, Examination Fever and Equality." In *Changes in the Japanese University, a Comparative Perspective,* edited by Walter K. Cummings, Ikuo Amano, and Kazaynki Kitamura. New York: Praeger Publishers, 1979.

Ferguson, Henry. *Manual for Multicultural Education.* Yarmouth, Maine: Intercultural Press, 1987.

Kidder, Louise H. "Requirements for Being 'Japanese.'" *International Journal of Intercultural Relations* 16 (1992): 383–93.

Suro, Roberto. "Study of Immigrants Finds Asians at Top in Science and Medicine." *Washington Post (*18 April 1994): sec. 6A.

Creating a Classroom Culture

A Pacific Perspective

8

Barbara J. Dougherty

Hannah Slovin

Annette N. Matsumoto

"Our problem is ready," said Ikaika. The teacher walked over to his group. "Everyone did their homework last night. But we liked Simeni's method so we're using that one."

The teacher asked, "Did everyone share a solution?"

"Yeah. But I did it the hard way. I liked how Simeni explained it," said Corey.

"We all voted for Simeni's way," said Sipa.

This conversation took place in an eighth-grade algebra class at the University Laboratory School (ULS), part of the Curriculum Research & Development Group, University of Hawaii. The school is our initial site for curriculum development, research, and evaluation activities.

ULS is specially designed for curriculum development for the Pacific region and the U.S. mainland. Three factors are considered so that ULS students are representative of Hawaii's population with regard to ethnicity, socioeconomic status, and achievement. ULS students, grades K–12, represent the State of Hawaii's multicultural population: Hawaiian, Japanese, Caucasian, Chinese, Samoan, Micronesian, Filipino, Korean, African American, and mixed ethnicities, as well as recent immigrants and citizens of foreign countries. They live in urban, suburban, and rural areas on the island of Oahu. They are from families of professional, semi-professional, skilled and unskilled workers, as well as from families where the parents are unemployed. Academic performance of middle-grades students range from the 5th to the 99th percentile on a standardized test. Students are grouped heterogeneously by academic performance, ethnicity, and socioeconomic backgrounds.

The school creates this heterogeneous environment so that any curricula developed there are applicable to a diverse student population. In Hawaii and the Pacific region, diversity in the classroom presents many challenges for teachers. Languages, traditions, and environments are more different than alike. Although ULS cannot be representative of all Pacific environments, it does provide a basis for curriculum development that is more applicable to a broader range of students.

Two projects from the CRDG-ULS site have produced student and teacher materials, instructional strategies, and assessment techniques specifically for classes with such diverse students. The Hawaii Algebra Learning Project (HALP) and Reshaping Mathematics Project (RMP) have developed curricula that improve student achievement for *all* students, regardless of cultural and social factors. Together they form the mathematics program for grades 6–8.

Although these curricula differ in content, they share at least four common characteristics necessary when high achievement is a goal for all students. These include—

- the use of research findings on student thinking and learning;
- the belief that students construct mathematical knowledge;

• multiple opportunities for student interaction;

• the allowance of time for concept development.

Most important, the curricula focus on using the diverse cultural and social backgrounds and experiences of the students to create a culture unique to each class. The culture is developed and negotiated by engaging in worthwhile tasks, establishing a safe, productive classroom environment, and creating opportunities for student interactions. Tasks, environment, and interaction overlap in the development of the classroom culture, but we present them separately to highlight some examples from our curricula.

WORTHWHILE MATHEMATICAL TASKS

Tasks play a major part in establishing and maintaining a learning environment where students share in the responsibility of shaping their own learning. The *Professional Standards for Teaching Mathematics* (NCTM 1991) recommends that teachers pose tasks based on "sound and significant mathematics; knowledge of students' understandings, interests, and experiences; and knowledge of the range of ways that diverse students learn mathematics." These tasks should engage students, develop their mathematical understanding, promote discussion, and draw on their diverse experiences.

In both the RMP and the HALP curricula, carefully crafted tasks that involve nonroutine and higher-order thinking introduce mathematical concepts. This is quite a change from a typical development. Over three to eight days, the tasks qualitatively change from problem solving to concept development and finally to skill, allowing students time to develop a deep understanding as they move to an algorithmic or skill level.

In order to draw on students' experiences, engage them, and develop their mathematical understanding while accommodating all students in our diverse population, we use open-ended tasks for at least three reasons. First, they have many possible solutions or have multiple solution paths. Students must pull from their experiences to create solutions that are often unique to individual cultures or experiences but are valued in developing the mathematics. Encouraging students to find as many solutions or solution strategies as possible respects multiple perspectives.

Additionally, in many Pacific cultures, like the Marshall Islands, learning occurs naturally through hands-on experiences. By carefully selecting problems that have multiple strategies, students have opportunities to solve problems using concrete materials as possible solution techniques.

Second, open-ended problems help students clarify concepts and make generalizations from the solutions presented during their discussions. These generalizations come from patterns that students notice in the answers or the solution strategies. Consistently asking students to describe patterns they notice within and across problems is an important component of instruction with Pacific Islanders because many of their cultural practices and traditions have long been based on patterns of some type.

Third, open-ended tasks provide a means for assessing student understanding. Consider the following problem given to our eighth graders (Rachlin, Matsumoto, and Wada 1992, p. 161):

Write an equation whose solution is 3.

Student responses in figure 8.1 describe four categories of student understanding: (1) equations without variables and 3 on the right side, (2) equations

Category	Student	Response
1	Jeff	6 − 3 = 3 (I did this because the question says to make an equation that equals 3.)
	Shaun	52 − 24 + 9 − 7 = 3
	William	1(2 + 1) = 3, 9x/3x = 3
2	Temujene	(9x + 3) ÷ 3 = 3
	Shalei	4y ÷ 8 = 3 , y = 6
	Litiana	x + 1 = 3 (x = 2)
	Vien	2y + 1 = 3
	Stephanie	9/3 + x − x = 3 ⌄ 3 ⌄ 0
3	Cathy	4x − 9 = 3
	Ross	2y − 3 = 3, y = 3
4	Kevin	19x + 35 − 25x = 17
	Tuan	5 + x = 8
	Heather	2y + 4y = 18

Fig. 8.1. Student responses to "Write an equation whose solution is 3."

with 3 on the right side, but whose solution is not 3, (3) equations with 3 on the right side, whose solution is 9, and (4) equations whose solution is 3.

The teacher can quickly assess each student's perception of what a solution to an equation means. Students in categories 1 and 2 view the solution to an equation as the expression on the right side of the equal sign. Students in category 3 are not quite sure whether a solution of 3 is the expression on the right side of the equal sign or the value of the variable that makes the equation true; both possibilities are covered. Students in category 4 have a solid understanding of what a solution to an equation means. As students share their equations and discuss which equations satisfy the conditions of the problem, they develop a better understanding of what an equation's solution is.

By encouraging multiple solution methods, a "routine" task takes on a new dimension.

> What polynomial multiplied by $x - 4$ equals $x^3 - 64$?
> (Rachlin, Matsumoto and Wada 1992, p. 405)

Although many students perceive this as a division problem and use the division algorithm, others see it as a multiplication problem in which they need to find the missing factor. In one of our eighth-grade classes, a sharing of different solution strategies led the class to appreciate the connections among multiplication, division, and factoring. The group that led the discussion shared three ways they solved the problem. (See fig. 8.2.)

In this problem, Zachary first explained how he worked backwards using the vertical method for multiplication. Next, Alyson showed how she got the same answer by dividing. Then, Eseta explained how she got the answer using the "box" method.

The "box" or lattice method is a popular method with our students. Eseta started off with a two-by-two box. She placed x^3 in the upper-left cell and −64 in the lower-right cell. She labeled the left side and proceeded to complete the diagram. She realized that she needed to add a third column of cells so that the entries in that column would cancel out the entries $-4x^2$ and $16x$. Like Zachary and Alyson, Eseta concluded that $x^2 + 4x + 16$ was the answer.

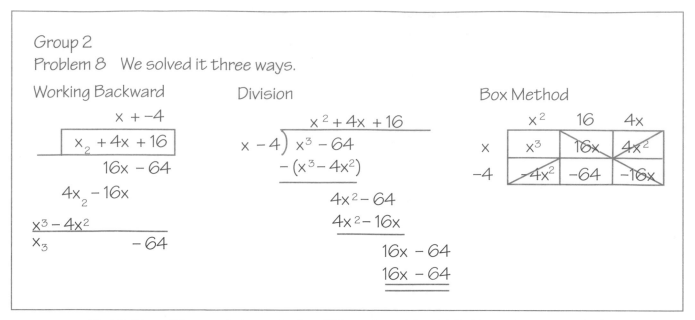

Fig. 8.2. Three different solution methods to "What polynomial multiplied by $x - 4$ equals $x^3 - 64$?"

Although this particular method may seem "nonalgebraic," it helps students organize their thinking. This structure helps students who lack confidence and experience in mathematical problem solving.

Most teachers say that their students do not like word problems because they lack reading comprehension skills or they do not have the skills to identify what arithmetic operations or algebraic techniques to use. Language skills are often the stumbling blocks in solving word problems. When students are not proficient in English, or when their primary language or cultural background does not include a particular idea, comprehension becomes an obstacle. For example, students from the Marshall Islands do not have the concept of large numbers or infinity because this idea is not represented in their language. Changing instructional approaches has helped students become more proficient with word problems, but another technique we use is to have students create their own word problems for a given equation.

Eighth graders at ULS were given the following writing prompt (Matsumoto, Dougherty, Wada, and Rachlin 1994, pp. 4–12):

Write a word problem that can be solved by the equation $4y - 7 = 3y$.

Students are also challenged because both sides of the equation contain variables. (See fig. 8.3.)

ENVIRONMENT

The classroom environment is central to student learning in our curricula. It "forms a hidden curriculum with messages about what counts in learning and doing mathematics" (NCTM 1991, p. 56). We view the classroom environment as one that melds the teacher's and students' beliefs about mathematical knowledge and what is considered reasonable mathematical activity. (Fieman-Nemser and Floden 1986; Lubinski 1994). This is more than physical space and materials; it establishes who is considered knowledgeable, by whom, how ideas are exchanged and disagreements resolved, and to what extent all students participate and learn in the class.

Mahea	Sandy had 4 purple balloons and y red balloons in his hands. S gave 7 of those balloons to a child. Now she only has three purple balloons and y red balloons. How many balloons did Sandy have (altogether)?
Eseta	You can't make a problem because $4y - 7$ you can't simplify and I think $3y$ isn't the answer.
Lesli	$4y - 7 = 3y$ $y = 7$ $28 - 7 = 21$ On the first day at the garage sale Kelly sold four items for y dollars and made 28 dollars. The next day she sold three items for y dollars. How much more money did Kelly make the first day?
Iris	Tom bought 4 boxes of candy right before Halloween. There was one package that was already opened so he got 7 dollars refund. So now he has 3 boxes of candy instead of 4 boxes to give to trick-or-treaters. Figure out the cost of each box.
Colin	George had 4 boxes of *Jurassic Park* books and sold 7 copies to the bookstore for $2 each. He is then left with 3 boxes of *Jurassic Park* books. So how many books are in one box?
Alyson	A football player ran the same amount of yards four times, and the fifth time lost 7 yards. Someone said that the player actually ran the amount of yards three times. How many yards did he run each time?
Sean	John had four times as many pencils as I have, then I broke 7 of his pencils in a pencil fight. Now he has only three times as many pencils as I have. How many pencils do I have?

Fig. 8.3. Students' journal responses to "Write a word problem that can be solved by the equation $4y - 7 = 3y$."

In our classes, the teacher's task is to understand students' mathematical thinking and to provide opportunities for the students to reshape it. This teaching role, given the diverse population that we have at the ULS, makes the classroom environment an essential component of success in teaching and learning mathematics.

When we try to implement a new teacher role, our perspective and the students' perspectives are often at odds. We may want to foster a learning environment that emphasizes conceptual understanding and mathematical processes; our students may want the instruction to concentrate on the rules and definitions used to solve exercises. In the students' minds, assignments are for the purpose of practicing skills and showing what one knows. Our conception and that of the National Council of Teachers of Mathematics (1991), nevertheless, is that mathematics is a body of knowledge to be explored; problem solving is the essence.

These differences in perceptions must be negotiated among the teacher and students. These negotiations often require an acclimation period of about six weeks before students accept the responsibility of discussing problems. But, they are necessary if classroom practices are to change.

With Pacific island students, the acclimation is slightly different because discussion does not come naturally. They are comfortable with silence and can wait for long periods of time without talking. Creating an interactive environment in this case requires detailed instructions so that they understand what discussions consist of and why they are necessary in a mathematics class.

AN ENVIRONMENT THAT SUPPORTS TEACHING AND LEARNING

Two related elements impact the relationship between the teacher's and the students' classroom roles: student responsibility and control of the level of mathematical content. Student responsibility is the capability of, and the opportunity for, students to think independently in the way they approach solving problems, justify their solution methods, and reflect on the reasonableness of their answers. We often expect students to organize themselves to work on tasks, select a place in the classroom to work, get needed resources, and find a partner when pairs are called for. We want students to monitor their progress on a task, ask for help as needed, and make adjustments when they find that their actions are not productive.

One of the key ways we support students' responsibility is by allowing them to comment on each other's problem solutions, ask questions, and suggest corrections. We are careful to wait after a student has presented a problem solution for others to initiate a discussion. Students usually begin questioning the presenter with questions such as "What method did you use?" or "How did you think to use that method?" Even though these questions are appropriate for the teacher to ask, they are much more powerful when they come from students.

Many of our tasks are designed to create controversy and bring out varying points of view about a particular concept. Where there are opposing points of view, however, we do not arbitrate for the students or try to force a consensus before they are ready to agree. Our role is to help students explore the different ways of looking at the problem within mathematical appropriateness. This facilitation role is particularly important when working with Pacific cultures because in many Pacific entities it is not appropriate to question others or to disagree face-to-face. We must be cautious so as not to push arguments to uncomfortable levels while we are encouraging students to speak up when they are not satisfied with an explanation or an idea.

Students are responsible for sharing their ideas and problem solutions in their group. As students discuss the problem they will present before the class, they are also accountable for questioning others. As students make presentations, others often take what is typically the teacher's role of questioning for justification and clarification. In such cases, we step back and allow the discussion to progress among the students. When one of our sixth-grade students, Nicole, was having difficulty describing how she solved a certain problem and the class could not see her point of view, the teacher prompted, "What might Nicole do to help you understand what she's saying?" "Show us [on the overhead]" and "Draw a picture" were the students' suggestions.

Encouraging students to do much of the questioning during presentations and discussions establishes norms and expectations within the learning environment. It is expected that students will listen to each other critically. It is not enough for them to sit passively during a presentation. As teachers, we participate in the discussion, but we do not raise all the questions for the students, nor do we automatically repeat what has been said. Otherwise, students learn to wait until the teacher repeats what the presenting student has said. This also makes the presenter accountable for expressing ideas clearly.

Student responsibility is central to the many collaboratively-organized tasks in our classrooms. As students work in groups, we monitor to ensure that all are participating. We recognize that the students may not have had experience working in groups and that it takes teacher effort to help students engage in successful group experiences. We foster the idea that a solution offered by a group's representative actually reflects the work and thinking of the entire group or that an individual has helped the group understand and accept his or her ideas.

As part of the students' responsibility, they must take the initiative to make judgments about their own work and that of others in relation to the mathematics. We want students to go beyond the criterion of whether the answers are correct or incorrect and assess the quality and quantity of the explanation. We believe that when students show each other how they solve the problems, it enables others to participate in the lessons. We often ask, "What did the presenter do to help you understand this problem?"

The second element that impacts the interaction between the teacher's and students' roles is who controls the level to which the mathematical content is pushed. We share control with the students by using prompts such as "Have you ever seen anything like this before?" and "How would you explain this?" These push students to share their own ideas and to use them to verify an answer. When students respond to these and other questions, we use their responses to either craft follow-up questions or to propose a twist on the problem. This sends the message that the students' thinking and understandings are valuable resources in constructing the mathematics.

We want students to value their own and each other's knowledge. Our role in classroom discourse is often that of a coach, helping students learn to communicate effectively with each other. We believe it is important for students to talk directly to each other, not through us. When a student has a comment about another student's work, we direct him or her to ask the other student, not us. It gives students a sense of mathematical power to be able to question another student, as well as a feeling of ownership of the work. During a group investigation, we might, for example, require that students check with other group members if they have a question, or we might stop the class and have them pose the question to the entire group.

ESTABLISHING A CLASSROOM ENVIRONMENT

The beginning days are important for setting the tone of the entire school year. We devote much time to establishing the culture of the classroom and reinforcing desirable behaviors. We give students tasks and problems to demonstrate that their experiences will be different from their other mathematics classes. At the same time, we are sensitive to students who are unfamiliar with this environment.

Some of the *Getting Started* problems and tasks from the RMP are designed to help students participate in class discussions. In one task, a student holds a card with an intricate design and tries to direct another student to re-create it by following oral directions. Students practice focusing on other students' descriptive language and explanations, relying on others for important information, and becoming less self-conscious when their responses are the focus of attention.

Trust building is important in setting the classroom environment. To become risk takers, students must feel that they will not be ridiculed when their answers are incorrect or their suggestions do not lead to a solution. We include tasks that may have more than one reasonable solution or solution strategy. Students learn that they can have a "different" answer than the rest of the class and still be doing productive work. In fact, we often see students valued as group members because they can think about problems in different ways.

To build the classroom environment, we assign tasks that illustrate student work expectations. Some early open-ended tasks are designed to introduce controversy. (See fig. 8.4) In the light problem, students often group solutions that are within the realm of possibility and solutions that are more fantastic. Other tasks are designed to prompt the use of specific vocabulary. The "J" problem raises questions about words like "same" and "different," and the discussion often involves clarifying terms that deal with position and orientation.

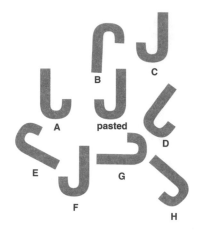

The Light Problem

Ben turned off the light in his bedroom and was able to get to bed before the room was dark. His bed is 15 feet from the wall switch. How did Ben do it?

The "J" Problem

Carrie was pasting some letter J's on a large poster (see sidebar). She had only gotten one done when a gust of wind scattered the others all around. What happened to all the J's? Write about the position of each letter J in relation to the one that was pasted on the poster.

Fig. 8.4. Tasks designed to introduce controversy and use subject-related terminology

Classroom Environment Reflects a View of Mathematics

Our perception of mathematics—a community activity where members create and justify solutions to problems—is reflected in the classroom environment. While we want the environment to support students taking risks, making conjectures, and "playing with ideas," we also want students to use some standard of sound mathematics as they express their ideas.

In the beginning of the year, we always follow a student's explanation with a request for justification or further explanation. Many students become annoyed or uncomfortable when they have difficulty talking about their thinking, and they may respond to our questions with "I don't know," "It just is like that," or "We learned it last year." As the year progresses, the students anticipate that we will ask for more justification or another student will ask for a more complete explanation. Justifying one's answers becomes an expected part of the learning environment.

We also want our students to reflect on their understandings with an analysis of the solutions resulting from a particular strategy. By giving students the opportunity to evaluate their work and that of others, they learn that their responsibility does not end with an answer. We believe that we are conveying the idea that their mathematical knowledge is valued and that they are capable of making decisions about appropriateness and correctness.

INTERACTIONS

In our culturally diverse classes, our focus is to develop rich interactions among students, with the teacher serving as facilitator and exploration guide. These interactions include problem discussions, questioning, and writing.

Problem Discussions

Problem discussions are in two forms: intragroup and intergroup interactions. These interactions contribute to the learning in both mathematical content and process.

Intragroup interactions focus on students sharing their ideas or solutions about homework problems, in-class problems, extensions of problems, or lab explorations. Our students are assigned less than eight homework problems every night to complete outside of class. The next day, in their small groups or pairs, they share their solutions to a problem that has been either assigned to or selected by the group. The goal of the group discussion is to prepare for a presentation to the class.

During this preparation time, students reach several decisions. The first is how the group will function. Students are not just cooperating to learn, but learning to cooperate (Good, Mulryan, and McCaslin 1992). With students from so many backgrounds, group functioning is a crucial piece of the classroom culture. No single method works for every group. Instead, we allow students to negotiate with group members on procedures they will follow and who will be responsible for what. They know the goal of the group's task and must decide how best to achieve it.

Second, groups must decide how to handle the mathematical content. As they share their ideas about a problem, they learn to disagree and offer constructive advice without offending or alienating group members. This requires that students listen critically and respond in a way that is viewed as helpful by group members as they assess the validity of the mathematics being presented. This is a dramatically different role for students who come from classrooms where the teacher is the only source of determining whether the mathematics is correct or not.

Finally, group members must be sensitive to the cultural diversity of the students in their group. For example, it is important in many cultures to "save face"—for someone not to feel like he or she is wrong. When students anticipate their answer may be seen as incorrect, they often avoid situations where they might be criticized. This affects the class contributions these students make, since, in their minds, it is better not to contribute at all than to risk criticism. When all students are sensitive to this cultural trait, they tend to respond with respect for, and positive recognition of, all the ideas presented.

After discussing the problems in groups, a student randomly selected from each group presents the problem to the class. Speaking before a peer group can be intimidating for middle school students. The class presentation, however, represents the collective thoughts of the group and lessens the presenter's apprehension. The group supports the presenter by interjecting, clarifying, or defending ideas when the presenter requests assistance.

The representation of group thinking in the problem presentations is a collaborative approach to learning and teaching, as shown by one of our students, Cathy, an eighth-grader. She responded to the journal prompt "How is this class different from other math classes?" and wrote

> In this class, it's collaboration. What I mean is, we really have like a community instead of a class. Our groups are like neighborhoods. We learn who we can trust, who we can depend on for what. And, we understand everybody in our group. Like, what they like and don't like, and how they work. Some people don't like to share first; they want to wait for somebody else to talk first. That's okay, we understand that. But, everybody knows they have to do their part. That's what neighborhoods do and that's what we do.

Cathy describes how collaborative groups accommodate individual differences while respecting the contributions each can make. Accommodation and respect start slowly at the beginning and build during the year, so that group interactions become qualitatively and quantitatively better over time.

Groups are an important part of developing good discussions about problems. They allow many students to be heard rather than just the one or two who would be brave enough to speak in front of the whole class. With a group's backing of a problem solution, the presenter can be more confident as he or she discusses the problem and its implications. Groups also help develop higher-level mathematics by allowing several students to put their thinking into one problem.

Questioning

Part of the interactions in our classes is motivated by teacher and student questions. Questions are invitations for interaction, and asking questions conveys the idea that you respect the knowledge, opinions, or ideas of others.

In our classes, we focus on problem-solving questions that require expanded responses. They cluster into three types—(1) reversibility, (2) flexibility, and (3) generalization (Krutetskii 1976). Each of these question types allows students to formulate and use novel, but mathematically appropriate, strategies in their solutions. This acknowledges and respects varying students' backgrounds and experiences. The level of the mathematics also adjusts to the level of the students because the questions require expanded explanations and foster multiple correct answers. This is an important aspect in classes that include such a diverse range of students.

The type of question asked is as important as who asks the question. At the beginning of the year, teachers model certain questions, such as "Can you solve the problem another way?" Students soon begin imitating the teacher in the problem presentations by asking the same question. At first, this seems to be superficial because students are merely repeating what the teacher has done. But, as the year progresses, these questions occur naturally in student discussions because students are genuinely interested in what other students did and why.

Who asks the questions is not the only change; there are changes in who answers the questions. If a student asks a question, we pause and wait for students to take the lead in answering it. Instead of students relying on the teacher, they now look to each other. Students take a greater interest in participating in the discussion because they have made the questions and the answers their own.

Through class interactions, explicit ideas about student and teacher roles begin to emerge. This is evident in this exchange that occurred in a class.

S_1: Dr. D, is this answer right?

S_2: Why are you asking her? She's just going to ask us another question. That's her job. Our job is to figure out if our answer is right or not.

T: What question do you think I would ask?

S_2: I think you would ask how we could justify the answer. If we can justify it, it's probably going to be right.

This dramatic shift in students taking responsibility for justifying and validating their solution processes and answers is an important part of the evolution of the classroom culture. Asking students to validate and justify communicates to the students that their reasoning is valued and expected in all interactions.

Additionally, when students must explain their thinking, it forces them to understand the mathematics at a deeper level.

Writing

Another interaction occurs when students write in our classes. Writing is done when students are solving homework problems, completing labs in class, responding to journal prompts (Dougherty 1996), or answering test items. Writing in multiple contexts extends the expectation that expanded responses are necessary and link students' spoken responses with written ones. Students know what is expected whether they are writing or speaking mathematics. It becomes second nature for them to give an explanation, whether they were explicitly asked to or not.

Writing also provides another means for students to share their work with others. It gives students time to digest and ponder what another student is saying. Unlike spoken language that goes at a fast pace, the written pieces can be reread and discussed until there is no misunderstanding about what a student is trying to communicate.

Using writing gives students who are uncomfortable speaking in front of their peer group the opportunity to communicate in another way. Many students who are not comfortable speaking in front of other students at the start of the school year are more adept at writing what they think. Writing tasks prevent students from feeling left out of discussions and indicate their ideas are valued.

Writing, however, also creates some difficulties for students. In Micronesia, the languages are primarily oral languages. When Pacific Islanders are confronted with writing tasks, it is often difficult for them to express themselves. Their writing requires immediate feedback and profits from the use of the writing process where revisions are expected.

Each of these forms of interactions is part of the evolution of the classroom environment. They enhance the respect for individuals in several dimensions, including culture, gender, and knowledge. By including these factors in building the classroom environment, a new culture is formed based on the contributions of both students and teachers as they develop mathematical understandings.

Implications

A culturally diverse class creates several implications for our teachers and students. Since 1983, when classes, teachers, and students have been observed and interviewed, at least six components to negotiate classroom culture have been identified. (See fig. 8.5 on the next page.)

First, our teachers and students truly believe that the construction of mathematical content is done in an environment where students are interacting within the classroom community. Students' ideas form new learnings that are comprised of multiple perspectives, tolerant of differences, and rich with cultural nuances.

Second, a safe environment is necessary if our students are to publicly share ideas. This doesn't mean that whatever students say will be validated—it means that students are not openly criticized and that incorrect answers are explored rather than tossed aside.

Third, multiple cultures in our classes require an instructional approach that allows and encourages students to analyze others' perspectives to see how they fit into the development of their own mathematics. The diversity of their ideas builds a respect between and among students for different ways of thinking.

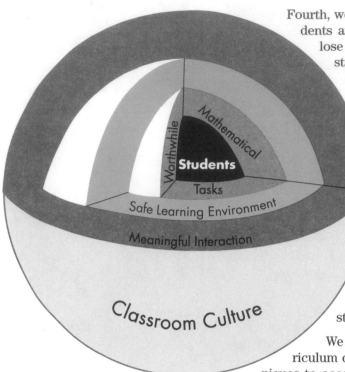

Fig. 8.5. Components of classroom culture

Fourth, we have found that we cannot ignore the fact that our students are a diverse group. By disregarding their diversity, we lose the richness of mathematical content and associated strategies and processes, thus detracting from student engagement.

Fifth, tasks that focus on worthwhile mathematics and *require* expanded responses drive the content in our classes. Without tasks that lend themselves to discussion, students have no varying approaches and thinking to share. Algorithmic-type problems are not enough. We have also found that these tasks must engage all students, not just appeal to a few.

Finally, our classes have shown us that we must focus on teaching students how to listen critically, modeling appropriate questioning techniques and maintaining consistent expectations of student responses. These factors create richer and more substantive interactions.

We are, however, left with a fundamental question. As curriculum developers, have we developed these instructional techniques to accommodate our students from diverse cultures OR have our instructional techniques been shaped by our students' multiple perspectives? Regardless of which is true, we know that we have experienced classes rich in mathematics, respectful of diversity, and abundant with student achievement.

REFERENCES

Dougherty, Barbara J. "The Write Way: A Look at Journal Writing in First-Year Algebra." *Mathematics Teacher* 89 (October 1996): 556–60.

Fieman-Nemser, Sharon, and Robert E. Floden. "The Cultures of Teaching." In *Handbook of Research on Teaching*, edited by M.C. Whittrock (pp. 505–26). London: Collier-Macmillan, 1986.

Good, Thomas L., Catherine Mulryan, and Mary McCaslin. "Grouping for Instruction in Mathematics: A Call for Programmatic Research on Small-Group Processes." In *Handbook of Research on Mathematics Teaching and Learning* edited by D.A. Grouws (pp. 165–96). New York: Macmillan Publishing Company, 1992.

Krutetskii, V. A. *The Psychology of Mathematical Abilities in School Children*, edited by J. Kilpatrick and I. Wirszup. Chicago, Ill.: University of Chicago, 1976.

Lubinski, Cheryl A. "The Influence of Teachers' Beliefs and Knowledge on Learning Environments." *Arithmetic Teacher* 41 (April 1994): 476–79.

Matsumoto, Annette N., Barbara J. Dougherty, Li Ann T. Wada, and Sidney L. Rachlin. *Algebra I: A Process Approach Teacher's Guide.* Honolulu, Hawaii: Curriculum Research & Development Group, University of Hawaii, 1994.

National Council of Teachers of Mathematics. *Professional Standards for Teaching Mathematics.* Reston, Va: National Council of Teachers of Mathematics, 1991.

Rachlin, Sidney L., Annette N. Matsumoto, and Li Ann T. Wada. *Algebra I: A Process Approach.* Honolulu, Hawaii: Curriculum Research & Development Group, University of Hawaii, 1992.